Lecture Notes in Biomathematics

Managing Editor: S. Levin

79

Przemyslaw Prusinkiewicz
James Hanan

Lindenmayer Systems, Fractals, and Plants

with Contributions by A. Lindenmayer, F. D. Fracchia, K. Krithivasan

Springer-Verlag
New York Berlin Heidelberg London Paris Tokyo Hong Kong

Authors

Przemyslaw Prusinkiewicz
James Hanan
Department of Computer Science
University of Regina
Regina, Saskatchewan, S4S 0A2 Canada

Mathematics Subject Classification (1980): Primary: 68U05,
Secondary: 68Q50, 92-04, 92A90

ISBN 3-540-97092-4 Springer-Verlag Berlin Heidelberg New York
ISBN 0-387-97092-4 Springer-Verlag New York Berlin Heidelberg

Printed in Germany

Printing and binding: Druckhaus Beltz, Hemsbach/Bergstr.
2849/3140-543210 – Printed on acid-free paper

Preface

L-systems are a mathematical formalism which was proposed by Aristid Lindenmayer in 1968 as a foundation for an axiomatic theory of development. The notion promptly attracted the attention of computer scientists, who investigated L-systems from the viewpoint of formal language theory. This theoretical line of research was pursued very actively in the seventies, resulting in over one thousand publications. A different research direction was taken in 1984 by Alvy Ray Smith, who proposed L-systems as a tool for synthesizing realistic images of plants and pointed out the relationship between L-systems and the concept of fractals introduced by Benoit Mandelbrot. The work by Smith inspired our studies of the application of L-systems to computer graphics. Originally, we were interested in two problems:

- Can L-systems be used as a realistic model of plant species found in nature?
- Can L-systems be applied to generate images of a wide class of fractals?

It turned out that both questions had affirmative answers. Subsequently we found that L-systems could be applied to other areas, such as the generation of tilings, reproduction of a geometric art form from East India, and synthesis of musical scores based on an interpretation of fractals.

This book collects our results related to the graphical applications of L-systems. It is a corrected version of the notes which we prepared for the ACM SIGGRAPH '88 course on fractals.

We would like to thank Aristid Lindenmayer for many stimulating discussions and guidance in the fascinating world of L-systems. Benoit Mandelbrot's appreciation of the relationship between L-systems and fractals gave us the motivation for surveying the results obtained so far. Kamala Krithivasan guided us through the intricate world of kolam patterns. Dave Fracchia implemented a program for modelling cell layers using map L-systems; an earlier implementation was kindly made available to us by Mark de Does. The map L-system data files used to produce images included in this book were prepared by Martin de Boer. Lynn Mercer wrote utilities for printing IRIS raster files on the laser printer. Also, we would like to thank Heinz-Otto Peitgen, Dietmar Saupe and Gerhard Rossbach, whose interest in our work made the original notes and this publication possible.

The support of our research by the Department of Computer Science, University of Regina, and the Natural Sciences and Engineering Research Council of Canada is gratefully acknowledged.

Przemyslaw Prusinkiewicz
James Hanan
Regina, March 1989

Sources

This book collects results from and includes edited parts of the following papers:

- P. Prusinkiewicz. Graphical Applications of L-systems. In *Proceedings of Graphics Interface '86 — Vision Interface '86*, pages 247–253, Vancouver, B.C., May 1986. [Included in Chapter 2 and Section 3.1]

- P. Prusinkiewicz. Applications of L-systems to computer imagery. In H. Ehrig, M. Nagl, A. Rosenfeld, and G. Rozenberg, editors, *Graph grammars and their application to computer science; Third International Workshop*, pages 534–548, Springer-Verlag, Berlin, 1987. Lecture Notes in Computer Science **291**. [Chapter 2, Section 3.1 and Section 3.6]

- P. Prusinkiewicz, A. Lindenmayer and J. Hanan. Developmental models of herbaceous plants for computer imagery purposes. Proceedings of *SIGGRAPH '88*, Atlanta, GA, August 1988. In *Computer Graphics*, 22:141–150, 1988. Used with the permission of the Association for Computing Machinery. [Sections 3.2, 3.3, 3.5. and Chapter 4]

- A. Lindenmayer and P. Prusinkiewicz. Developmental models of multicellular organisms: A computer graphics perspective. In C. Langton, editor, *Artificial life*, pages 221–249, Addison-Wesley, Redwood City, 1989. [Sections 1.2, 3.2, 3.3, 3.5, Chapter 4 and Chapter 5]

- P. Prusinkiewicz and K. Krithivasan. Algorithmic generation of South Indian folk art patterns. In *Proceedings of the International Conference on Computer Graphics ICONG'88*, Singapore, September 1988. [Section 6.2]

- P. Prusinkiewicz. Score generation with L-systems. In *Proceedings of the International Computer Music Conference '86*, pages 455–457, the Hague, the Netherlands, October 1986. [Section 6.3]

- A. Lindenmayer and P. Prusinkiewicz. An annotated bibliography on plant modelling and growth simulation. In C. Langton, editor, *Artificial life*, pages 625–643, Addison-Wesley, Redwood City, 1989. [Chapter 7]

Contents

Chapter 1

Introduction

Lindenmayer systems — or L-systems for short — find an increasing number of applications in computer graphics. Two principal areas include generation of fractals and realistic modelling of plants. More exotic applications, ranging from the reproduction of a traditional East Indian art form to graphically motivated algorithms for music composition, are also known. These notes survey various graphical applications of L-systems. A listing of a program which allows for the replication of most illustrations and for further experimentation with L-systems is included.

1.1 Rewriting systems

Before we proceed with technical details, let us focus on the concept of rewriting which is central to the notion of L-systems. The basic idea is to define complex objects by successively replacing parts of a simple initial object using a set of *rewriting rules* or *productions*. Usually the rewriting can be carried out recursively. The classic example of a graphical object defined in terms of rewriting rules is the "snowflake" curve (Figure 1.1), proposed in 1905 by von Koch [107]. Mandelbrot [72, page 39] restates this construction as follows:

Koch construction

> One begins with *two shapes*, an *initiator* and a *generator*. The latter is an oriented broken line made up of N equal sides of length r. Thus each stage of the construction begins with a broken line and consists in replacing each straight interval with a copy of the generator, reduced and displaced so as to have the same end points as those of the interval being displaced.

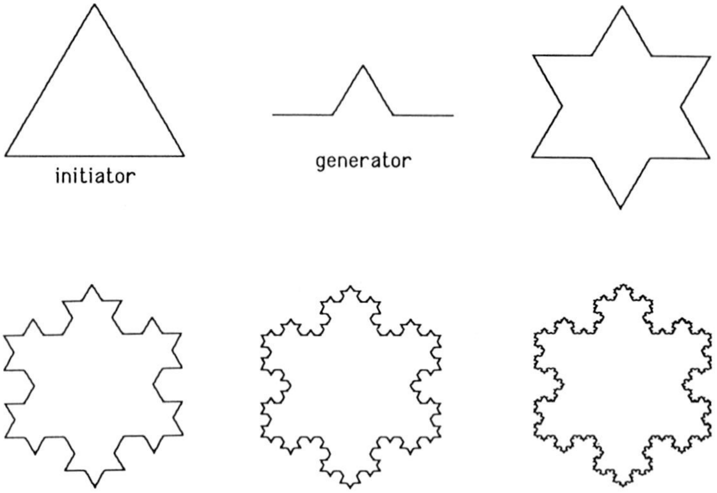

Figure 1.1: Construction of the "snowflake" curve.

While the Koch construction recursively replaces open polygons, rewriting systems which operate on other objects have also been investigated. For example, Wolfram [108,109] studied patterns generated by rewriting elements of rectangular arrays. A similar array-rewriting mechanism is the cornerstone of the popular game of life. An important body of research was devoted to various graph-rewriting systems [11,20,19].

Grammars

The most extensively studied and the best understood rewriting systems operate on character strings. The first formal definition of such a system was given at the beginning of this century by Thue (see [89]), but a wide interest in string rewriting was spawned in the late 1950's by Chomsky's work on formal grammars [10]. The concept of rewriting was applied there to describe syntactic features of natural languages. A few years later Backus and Naur introduced a rewriting-based notation in order to formally describe the programming language ALGOL-60 [4,77]. The equivalence of the Backus-Naur form (BNF) and the context-free class of Chomsky grammars was soon recognized [32] and a period of fascination with syntax, grammars and their application to computer science began. At the center of attention were sets of strings — called formal languages — and the methods for generating, recognizing and transforming them.

Languages and images

The advances in formal language theory brought up the idea of using character strings to describe pictures. The initial goal was to recognize objects, such as handwritten letters [76] or chromosomes [45,46], by coding

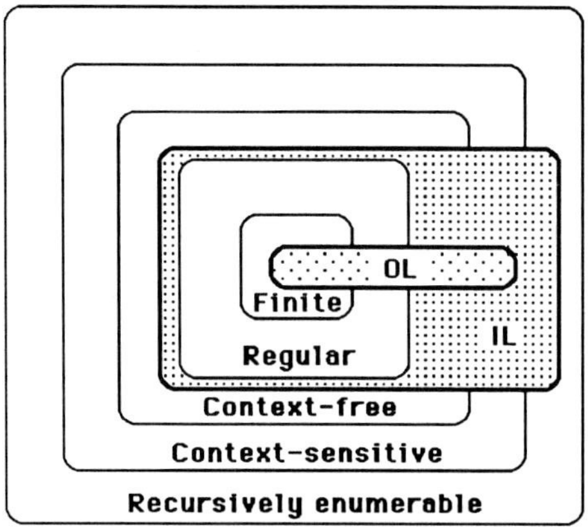

Figure 1.2: Relations between Chomsky classes of languages and language classes generated by L-systems. The symbols OL and IL denote language classes generated by context-free and context-sensitive L-systems, respectively.

their images, then analyzing the resulting strings. Among the variety of image representations considered, the chain coding proposed by Freeman [26] had a geometrically precise definition. Consequently, it was suitable for investigating the relationship between images and language classes in a formal way. Such a study was undertaken by Feder [23], (see also [31]), but the results were disappointing. They revealed that the languages corresponding to intuitively simple classes of images, such as straight lines of arbitrary slope, circles of arbitrary radius, and even rectangles in a square grid, were all context-sensitive. In contrast, most practically useful results of formal language theory referred to context-free languages. Thus, a discrepancy developed between the capabilities of the well understood class of languages and the requirements created by graphical applications.

L-systems

In 1968 a biologist, Aristid Lindenmayer, introduced a new type of string-rewriting mechanism, subsequently termed L-systems [54]. The essential difference between Chomsky grammars and L-systems lies in the method of applying productions. In Chomsky grammars productions are applied sequentially, one at a time, whereas in L-systems they are applied in parallel and simultaneously replace all letters in a given word. This difference has an essential impact on the properties of L-systems. For example, there are languages which can be generated by context-free L-systems (called OL-systems) but can not be generated by context-free Chomsky grammars (Figure 1.2).

Plant models

The introduction of L-systems revived interest in the representation of images using string characters. L-systems were conceived as a formal model of plant development, so initial efforts concentrated on the automatic generation of plant images. The first plots were published in 1974 by Frijters and Lindenmayer [30], and Hogeweg and Hosper [38]. In both cases, L-systems were used primarily to determine the branching topology of the modelled plants. The geometric aspects, such as the lengths of line segments and the branching angles, were added in a post-processing phase. The results of Hogeweg and Hosper were subsequently extended by Smith [94,98], who demonstrated the potential of L-systems in realistic image synthesis.

Fractals

A different approach to L-system interpretation was proposed in 1979 by Szilard and Quinton [101]. They concentrated on image representations with rigorously defined geometry (such as chain coding) and showed that strikingly simple context-free L-systems could generate the intriguing, convoluted curves known today as fractals. The results of Szilard and Quinton were subsequently extended in several directions. Siromoney and Subramanian specified L-systems which generated classic space-filling curves [93]. Dekking investigated the limit properties of curves generated by L-systems [18], and concentrated on the problem of determining the fractal (or Hausdorff) dimension of the limit set [17]. Prusinkiewicz presented more examples of fractals and plant-like structures modelled using L-systems; they were obtained using an interpretation based on a LOGO-like turtle [82] and its three-dimensional counterpart [81].

L-systems and graphics

Why are L-systems so successful in image generation applications? The answer seems to be directly related to the parallel operation of L-systems. On one hand, the parallelism links L-systems to the Koch construction of fractals. On the other hand, living organisms also develop in a parallel manner, so parallelism is an important factor in the simulation and modelling of plants. In the following chapters we survey several classes of graphical objects generated using L-systems. We hope that the mathematical beauty of the notion of L-systems, the simplicity of L-system-based algorithms for image generation and the intriguing images which can be obtained will encourage the reader to experiment further and extend the methods discussed.

1.2 DOL-systems

In this section we introduce the simplest class of L-systems, called DOL-systems. We start from an example which is intended to provide an intuitive

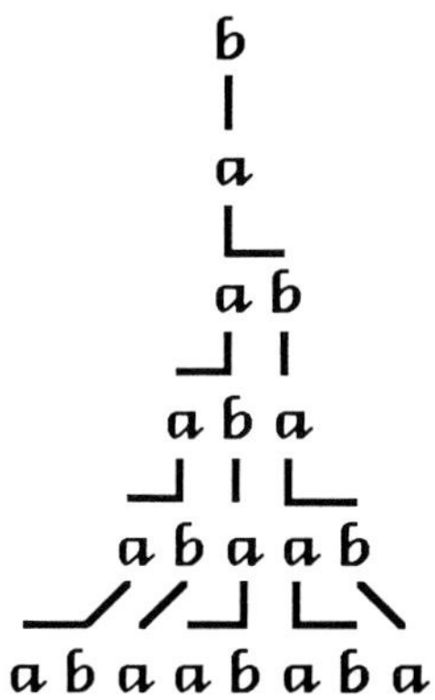

Figure 1.3: Example of a derivation in a DOL-system.

understanding of the main idea.

Example

Let us consider strings built of two letters a and b (they may occur many times in a string). For each letter we specify a rewriting rule. The rule $a \rightarrow ab$ means that the letter a is to be replaced by the string ab, and the rule $b \rightarrow a$ means that the letter b is to be replaced by a. The rewriting process starts from a distinguished string called the axiom. Let us assume that it consists of a single letter b. In the first derivation step (the first step of rewriting) the axiom b is replaced by a using production $b \rightarrow a$. In the second step a is replaced by ab using production $a \rightarrow ab$. The word ab consists of two letters, both of which are *simultaneously* replaced in the next derivation step. Thus, a is replaced by ab, b is replaced by a, and the string aba results. In a similar way (by the simultaneous replacement of all letters), the string aba yields $abaab$ which in turn yields $abaababa$, then $abaababaabaab$, and so on (Figure 1.3).

Below we present formal definitions describing DOL-systems and their operation. For more details see [37,88].

L-system

Let V denote an alphabet, V^* the set of all words over V, and V^+ the set of all nonempty words over V. A *string OL-system* is an ordered triplet $G = \langle V, \omega, P \rangle$ where V is the *alphabet* of the system, $\omega \in V^+$ is a nonempty word called the *axiom* and $P \subset V \times V^*$ is a finite *set of productions*. If a pair (a, χ) is a production, we write $a \rightarrow \chi$. The letter a and the word χ are called the *predecessor* and the *successor* of this production, respectively. It is assumed that for any letter $a \in V$, there is at least one word $\chi \in V^*$ such that $a \rightarrow \chi$. If no production is explicitly specified for a given predecessor $a \in V$, we assume that the *identity production* $a \rightarrow a$ belongs to the set of productions P. A OL-system is *deterministic* (noted *DOL-system*) if and only if for each $a \in V$ there is exactly one $\chi \in V^*$ such that $a \rightarrow \chi$.

Derivation

Let $\mu = a_1 \ldots a_m$ be an arbitrary word over V. We will say that the word $\nu = \chi_1 \ldots \chi_m \in V^*$ is *directly derived* from (or *generated* by) μ and write $\mu \Rightarrow \nu$ if and only if $a_i \rightarrow \chi_i$ for all $i = 1, \ldots, m$. A word ν is generated by G in a derivation of *length* n if there exists a *developmental sequence* of words $\mu_0, \mu_1, \ldots, \mu_n$ such that $\mu_0 = \omega$, $\mu_n = \nu$ and $\mu_0 \Rightarrow \mu_1 \Rightarrow \ldots \Rightarrow \mu_n$.

Anabaena

We will illustrate the operation of DOL-systems by one more example. The formalism is used here to simulate the development of a multicellular filament such as that found in blue-green bacteria *Anabaena catenula* and various algae [74,52]. The symbols a and b represent cytological states of the cells (these states have to do with their size and readiness to divide). The subscripts l and r indicate cell polarity, specifying the positions in which daughter cells of type a and b will be produced. The development is governed by the following rules:

$$\begin{aligned} p_1 &: \; a_r \rightarrow a_l b_r \\ p_2 &: \; a_l \rightarrow b_l a_r \\ p_3 &: \; b_r \rightarrow a_r \\ p_4 &: \; b_l \rightarrow a_l \end{aligned}$$

Starting from a single cell a_r (the axiom), the above L-system generates the following sequence of words:

$$\begin{array}{l} a_r \\ a_l b_r \\ b_l a_r a_r \\ a_l a_l b_r a_l b_r \\ b_l a_r b_l a_r a_r b_l a_r a_r \\ \ldots \end{array}$$

Under a microscope, the cells appear as cylinders of various lengths. The a-type cells are longer than the b-type cells. The corresponding schematic image of filament development is shown in Figure 1.4. Note that due to the discrete nature of L-systems the continuous growth of cells between subdivisions is not reflected in this model.

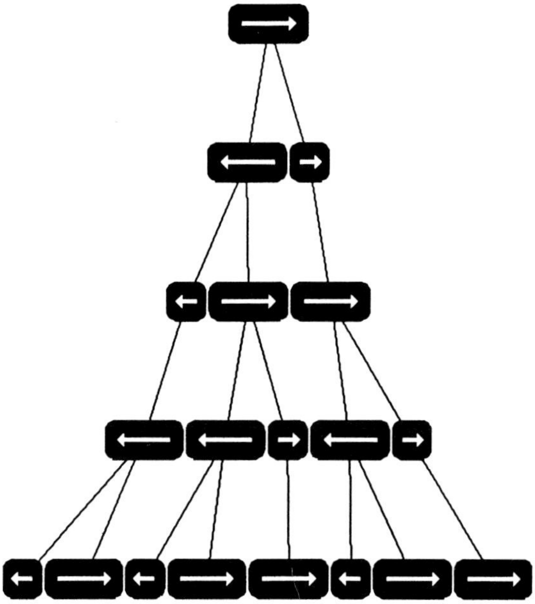

Figure 1.4: Development of a filament (*Anabaena catenula*) simulated using a DOL-system. Arrows indicate cell polarities.

Chapter 2

Fractals

The geometric interpretation of strings used in the previous chapter to generate schematic images of *Anabaena catenula* was a very straightforward one. Letters of the L-system alphabet were represented graphically as shorter or longer rectangles with rounded corners. The generated structures were essentially one-dimensional chains of rectangles reflecting the sequence of symbols in the corresponding strings.

Many fractals (or at least their finite approximations) can also be thought of as sequences of primitive elements — line segments. However, the lengths of segments and the angles between them play a crucial role. To produce fractals, strings generated by L-systems must contain the necessary information about figure geometry. We describe here a graphical interpretation of strings based on the notion of a LOGO-style *turtle* [1]. This interpretation was originally proposed by Szilard and Quinton [101].

Turtle

A *state* of the turtle is defined as a triplet (x, y, α), where the Cartesian coordinates (x, y) represent the turtle's *position* and the angle α, called the turtle's *heading*, is interpreted as the direction in which the turtle is facing. Given the *step size* d and the *angle increment* δ, the turtle can respond to commands represented by the following symbols (Figure 2.1a):

F Move forward a step of length d. The state of the turtle changes to (x', y', α), where $x' = x + d\cos\alpha$ and $y' = y + d\sin\alpha$. A line segment between points (x, y) and (x', y') is drawn.

f Move forward a step of length d without drawing a line.

\+ Turn right by angle δ. The next state of the turtle is $(x, y, \alpha + \delta)$. It is assumed here that the positive orientation of angles

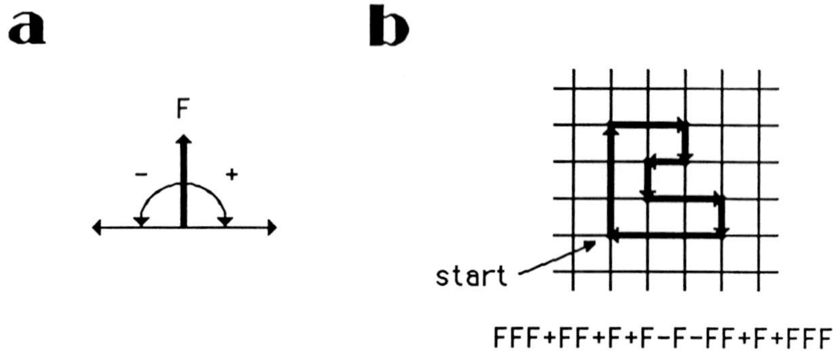

Figure 2.1: (a) Turtle interpretation of string symbols F, $+$, $-$. (b) Interpretation of a string. The angle increment δ is equal to 90°. Initially the turtle faces up.

is clockwise. (In some examples included in these notes the interpretation of the symbols $+$ and $-$ is inverted.)

$-$ Turn left by angle δ. The next state of the turtle is $(x, y, \alpha - \delta)$.

All other symbols are ignored by the turtle (the turtle preserves its current state).

Interpretation

Let ν be a string, (x_0, y_0, α_0) the initial state of the turtle, and d, δ fixed parameters. The picture (set of lines) drawn by the turtle responding to the string ν is called the ***turtle interpretation*** of ν (Figure 2.1b).

Having a rigorous method for mapping strings into pictures, we may apply it to interpret strings which are generated by L-systems. For example, Figure 2.2 presents four approximations of the "quadratic Koch island" taken from [72, page 51]. These figures were obtained by interpreting strings generated by the following L-system:

$$\begin{aligned} \omega &: \; F + F + F + F \\ p &: \; F \rightarrow F + F - F - FF + F + F - F \end{aligned}$$

The images correspond to the strings obtained in derivations of length 0–3. The angle increment δ is equal to 90°. The step length d is decreased four times between subsequent images. This makes the distance between the endpoints of the polygon represented by the production successor equal to the length of the line represented by the predecessor.

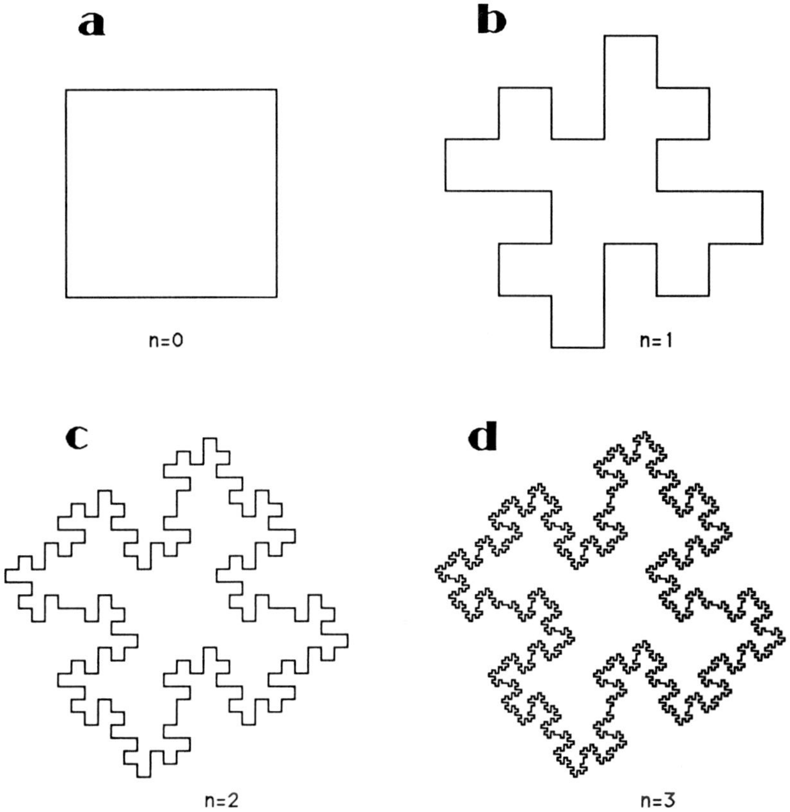

Figure 2.2: Generating a quadratic Koch island.

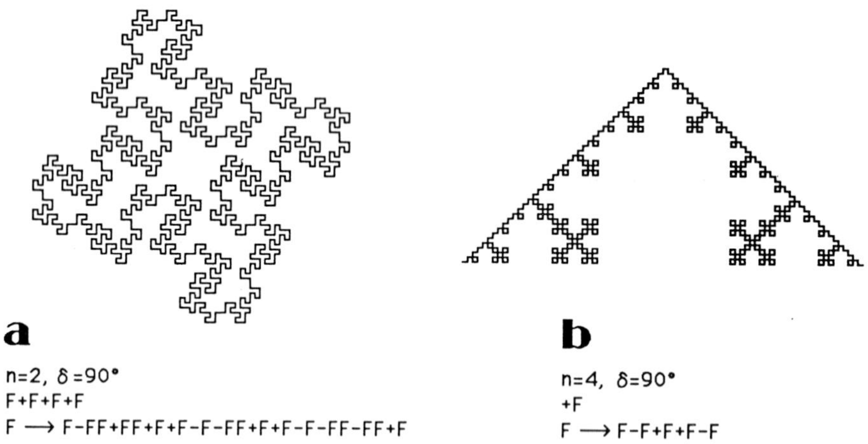

Figure 2.3: Examples of L-systems generating Koch curves. (a) Quadratic Koch island [Mandelbrot 1982, p. 52]. (b) A quadratic modification of the snowflake curve [Mandelbrot 1982, p. 139].

L-systems vs. Koch constructions

The above example reveals a close correspondence between Koch constructions (with a single generator) and L-systems. The initiator corresponds to the axiom, while the generator is represented by the successor of a single production; its predecessor is equal to F. L-systems specified this way can be perceived as *codings* of Koch constructions. Figure 2.3 presents further examples of Koch curves generated using L-systems. A slight complication occurs if the fractal is not connected; the second production (with the predecessor f) is then required to keep components the proper distance from each other (Figure 2.4). The ease of modifying L-systems makes them suitable for developing new Koch curves; for example, one can start from a particular L-system and observe the results of inserting, deleting or replacing some symbols. Some fractals obtained this way are shown in Figure 2.5.

L-system transformations

According to the original description of the Koch construction, the generator can be translated, rotated and scaled to match the original position of the replaced segment. More curves can be constructed if reflection is allowed as well. For example, the dragon curve shown in Figure 2.6 [72,15] can be generated by repetitively substituting line segments by pairs of lines forming either a left or a right turn. This is described by the following L-system:

$$\begin{aligned} \omega &: F_l \\ p_1 &: F_l \rightarrow F_l + F_r + \\ p_2 &: F_r \rightarrow -F_l - F_r \end{aligned}$$

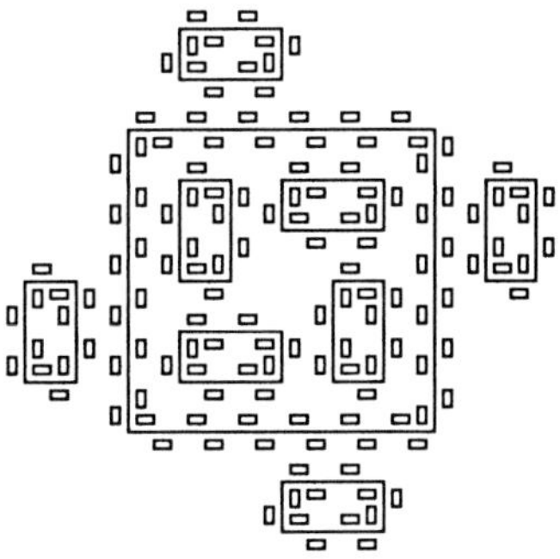

n=2, δ=90°
F-F-F-F
F ⟶ F-f+FF-F-FF-Ff-FF+f-FF+F+FF+Ff+FFF
f ⟶ ffffff

Figure 2.4: Combination of islands and lakes [Mandelbrot 1982, p. 121].

Two different symbols, F_l and F_r, are interpreted by the turtle as the "move forward" command. This is inconsistent with the original specification of turtle interpretaton. Fortunately, the use of many symbols with the same interpretation can be avoided by transforming the L-system under consideration. First, let us temporarily assume that the predecessor of a production can contain more than one letter; thus an entire subword can be replaced by the successor of a single production. The dragon-generating L-system can then be rewritten as:

$$\begin{aligned} \omega &: \quad Fl \\ p_1 &: \quad Fl \rightarrow Fl + rF+ \\ p_2 &: \quad rF \rightarrow -Fl - rF \end{aligned}$$

where the symbols l and r are not interpreted by the turtle. Then, notice that the production $Fl \rightarrow Fl + rF+$ replaces the letter l by the string $l + rF+$ while the leading letter F is left intact. In a similar way, the production $rF \rightarrow -Fl - rF$ replaces the letter r by the string $-Fl - r$ and leaves the trailing F intact. Thus, the L-system can be transformed into the final form:

$$\begin{aligned} \omega &: \quad Fl \\ p_1 &: \quad l \rightarrow l + rF+ \\ p_2 &: \quad r \rightarrow -Fl - r \end{aligned}$$

Now both productions have single-letter predecessors and the unnecessary use of several symbols with the same graphical interpretation has been avoided.

Space-filling curves

Figure 2.7 presents more examples of fractal curves constructed from "left" and "right" elements. The underlying L-systems have a similar structure to that of the dragon curve.

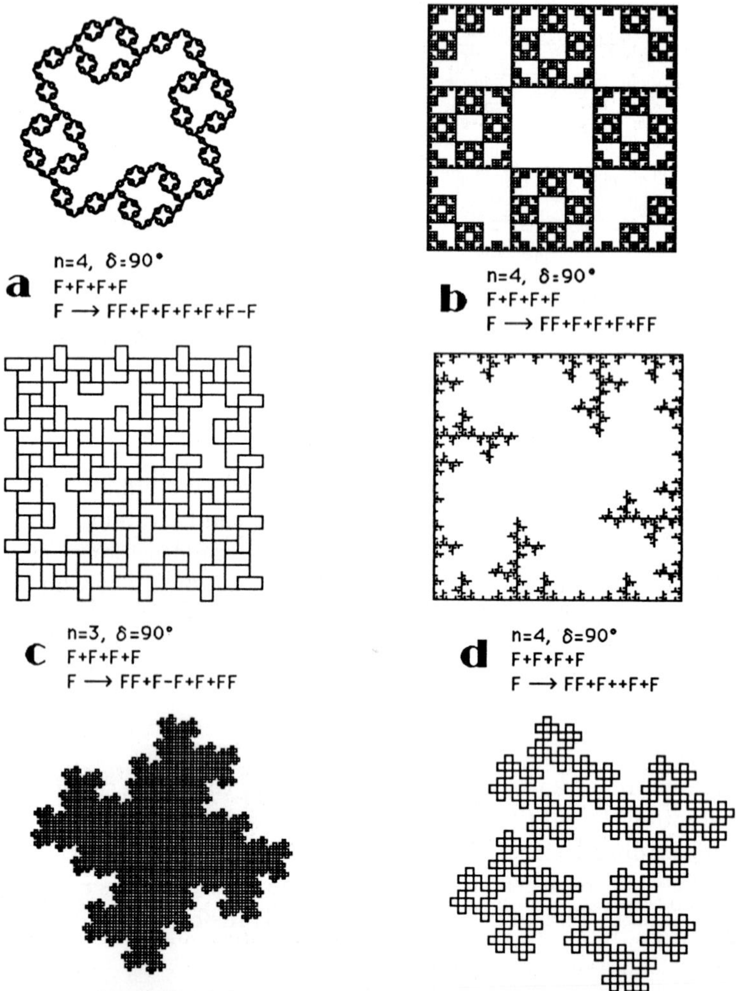

Figure 2.5: A sequence of Koch curves obtained by successively modifying the production successor.

Figure 2.6: The dragon curve [Davis and Knuth 1970].

In the examples discussed so far the symbols + and − were interpreted as a right and a left turn of 90°, respectively. Other values of angle increment δ can also lead to interesting images. For example, Figure 2.8 presents two fractal curves obtained using $\delta = 60°$.

The turtle interpretation of strings can be extended in many directions. Some extensions are presented below. Those pertinent to the modelling of plants are introduced in the next chapter.

Polygon filling

In its motion, a turtle may trace a closed non-self-intersecting curve (Jordan curve). Such a curve can be considered as the boundary of a polygon. We introduce two additional symbols, the opening brace { and the closing brace }, to delimit the substring which determines the boundary of a filled polygon. An example of a fractal with a filled polygon is shown in Figure 2.9.

Spline interpolation

Consecutive positions may be considered as control points specifying a smooth interpolating curve; for example a B-spline curve [24] (Figure 2.10). Such a modification yields pleasant visual effects and reveals the internal structure of the fractal better than the usual straight-line interpretation (Figure 2.11).

Turtle in 3D

The concept of turtle geometry can be extended to three dimensions [1]. Accordingly, we may introduce a three-dimensional turtle interpretation of strings. The key concept is to represent the current *orientation* of the turtle in space by three vectors $\vec{H}, \vec{L}, \vec{U}$, indicating the turtle's *heading*, the direction to the *left*, and the direction *up*. These vectors have unit length, are perpendicular to each other, and satisfy the equation $\vec{H} \times \vec{L} = \vec{U}$. Rotations of the turtle can then be expressed by the equation:

$$\left[\vec{H'} \;\; \vec{L'} \;\; \vec{U'} \right] = \left[\vec{H} \;\; \vec{L} \;\; \vec{U} \right] \mathbf{R}$$

where $\mathbf{R}$ is a 3×3 rotation matrix [24]. Specifically, rotations by angle α

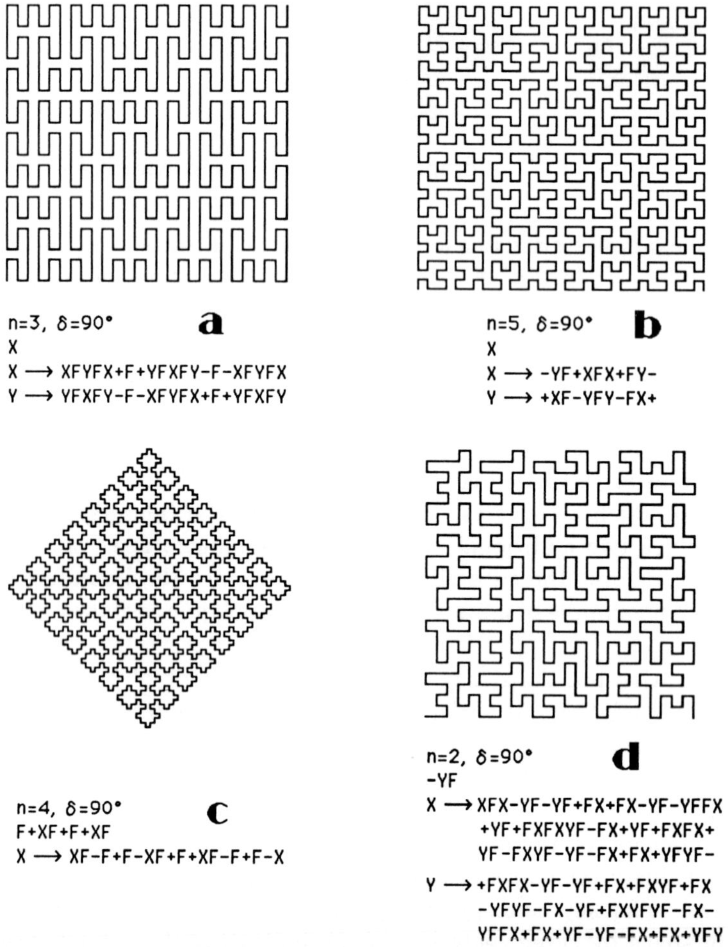

Figure 2.7: "Classic" space-filling curves and the corresponding L-systems. (a) Peano [1890] curve, (b) Hilbert [1891] curve, (c) A square-grid approximation of the Sierpiński [1912] curve, (d) Quadratic Gosper curve [Dekking 1982].

a n=6, δ=60°
YF
X ⟶ YF+XF+Y
Y ⟶ XF-YF-X

b n=4, δ=60°
XF
X ⟶ X+YF++YF-FX--FXFX-YF+
Y ⟶ -FX+YFYF++YF+FX--FX-Y

Figure 2.8: Examples of curves obtained using angle increment $\delta = 60°$. (a) Sierpiński arrowhead [Mandelbrot 1982, p. 142], (b) Hexagonal Gosper curve [Mandelbrot 1982, p. 70].

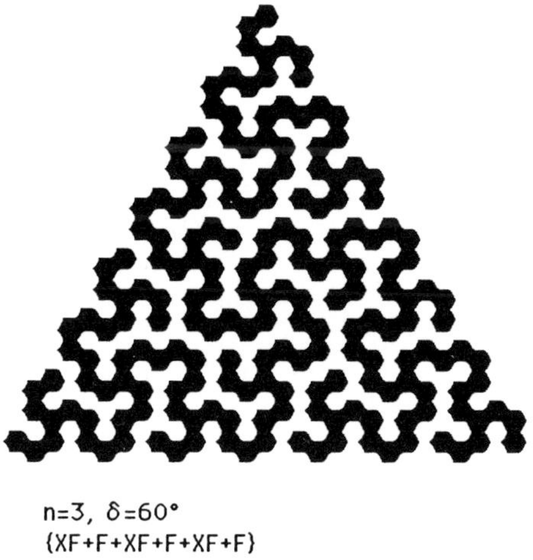

n=3, δ=60°
{XF+F+XF+F+XF+F}
X ⟶ XF+F+XF-F-F-XF-F+F+F-F+F+F-X

Figure 2.9: A fractal with a filled polygon [Szillard and Quinton 1979].

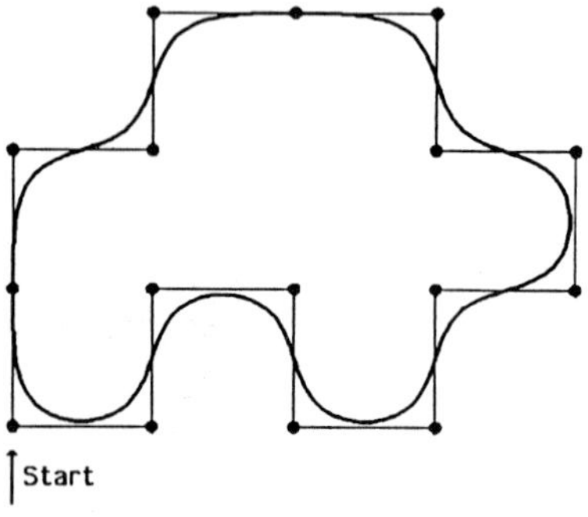

FF+F-F+FF+F-F+F+F-F+F+F-F-F+F

Figure 2.10: Turtle interpretation of a string with B-spline interpolation.

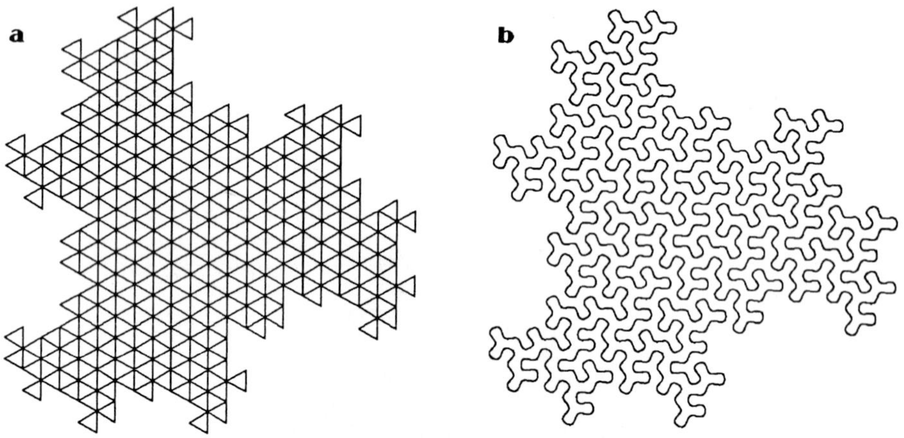

n = 3, $\delta = 120°$
F+F+F
F $\rightarrow$ F-F+F

Figure 2.11: A comparison of fractals obtained using (a) straight-line turtle interpretation [Dekking 1982] and (b) B-spline interpolation. The interpolated curve does not self-intersect and therefore represents the path of the turtle in a more clear way.

about vectors $\vec{H}, \vec{L}$ and $\vec{U}$ are represented by the matrices:

$$\mathbf{R_U}(\alpha) = \begin{bmatrix} \cos\alpha & \sin\alpha & 0 \\ -\sin\alpha & \cos\alpha & 0 \\ 0 & 0 & 1 \end{bmatrix}$$

$$\mathbf{R_L}(\alpha) = \begin{bmatrix} \cos\alpha & 0 & -\sin\alpha \\ 0 & 1 & 0 \\ \sin\alpha & 0 & \cos\alpha \end{bmatrix}$$

$$\mathbf{R_H}(\alpha) = \begin{bmatrix} 1 & 0 & 0 \\ 0 & \cos\alpha & -\sin\alpha \\ 0 & \sin\alpha & \cos\alpha \end{bmatrix}$$

The following symbols control turtle orientation in space (Figure 2.12):

+ Turn left by angle δ. The rotation matrix is equal to $\mathbf{R_U}(\delta)$.

− Turn right by angle δ. The rotation matrix is equal to $\mathbf{R_U}(-\delta)$.

& Pitch down by angle δ. The rotation matrix is equal to $\mathbf{R_L}(\delta)$.

∧ Pitch up by angle δ. The rotation matrix is equal to $\mathbf{R_L}(-\delta)$.

\ Roll left by angle δ. The rotation matrix is equal to $\mathbf{R_H}(\delta)$.

/ Roll right by angle δ. The rotation matrix is equal to $\mathbf{R_H}(-\delta)$.

| Turn around. The rotation matrix is equal to $\mathbf{R_U}(180°)$.

An example of a three-dimensional object created using an L-system is shown in Figure 2.13.

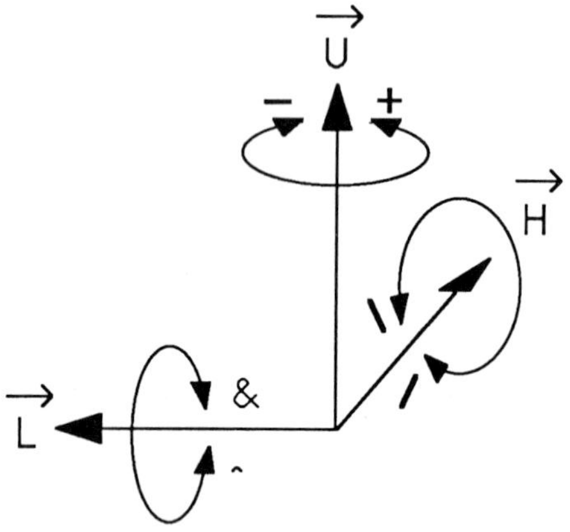

Figure 2.12: Controlling the turtle in three dimensions.

n = 2, $\delta = 90°$
&-XS-XS-XS-XS
X → XS+XS-XS-XSXS+XS+XS-X
S → F ∧ F ∧ F ∧ F ∧ F

Figure 2.13: A "paper-tape" version of the quadratic Koch curve.

Chapter 3

Models of plant architecture

3.1 Bracketed L-systems

According to the rules introduced in the previous chapter the turtle interprets a character string as a sequence of line segments, connected "head to tail" with each other. Depending on the segment lengths and the angles between them, the resulting line self-intersecting or not, can be more or less convoluted, with some segments drawn many times and others made invisible, but it always remains just a single line.

In his 1968 paper [54], Lindenmayer introduced a notation for representing graph-theoretic trees using strings with ***brackets***. The motivation was to formally describe branching structures found in many plants, from algae to trees, using the general framework of L-systems. Subsequently, geometric interpretations of L-systems operating on strings with brackets were introduced for the purpose of presenting modelled structures in the form of computer-generated plots [38,30] and realistic images [98]. An extension of turtle interpretation to bracketed strings and L-systems [82,81] is described below.

We introduce two new symbols interpreted by the turtle:

Stack operations

[Push the current state of the turtle onto a pushdown stack. The information saved on the stack contains the turtle's position and orientation, as well as other attributes such as the color and width of lines being drawn.

] Pop a state from the stack and make it the current state of the turtle. No line is drawn, although in general the position of the turtle changes.

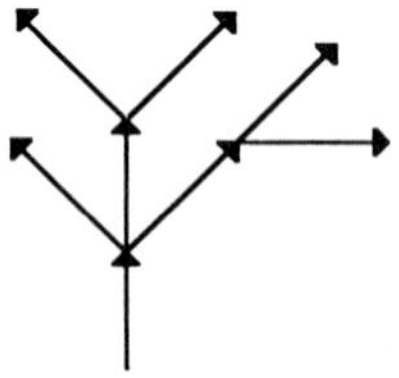

F[-F][+F[+F]F]F[-F][+F]

Figure 3.1: Turtle interpretation of a bracketed string.

An example of a bracketed string and its turtle interpretation are shown in Figure 3.1.

Derivations in bracketed OL-systems proceed as in OL-systems without brackets, the brackets being rewritten into themselves. Examples of two-dimensional branching structures generated by bracketed OL-systems are shown in Figure 3.2.

Bush Figure 3.3 is an example of a three-dimensional bush-like structure generated by a bracketed L-system. In order to make this L-system easier to analyze, identifiers are used instead of single-letter, non-interpreted symbols.

ω : **apex**
p_1 : **apex** → [**branch** //////' **branch** ///////' **branch**]
p_2 : **branch** → [& **internode leaf** ! **apex**]
p_3 : **internode** → *F* **leaf**
p_4 : *F* → *F* ///// **internode**
p_5 : **leaf** → [' ' ' $\wedge\wedge$ $\{-f+f+f-|-f+f+f\}$]

The angle increment δ is equal to 22.5°. The system operates as follows. Production p_1 creates three new branches from an apex of the old branch. Production p_2 describes a branch as consisting of an internode, a leaf and an apex (which will subsequently create three new branches). Productions p_3 and p_4 specify internode growth. In subsequent derivation steps the internode gets longer and acquires new leaves. This violates a biological rule of *subapical growth* (discussed later), but produces an acceptable visual effect in a still picture. Production p_5 specifies the leaf as a filled polygon with six edges. The symbols ! and ' are used to decrement the diameter of segments and increment the current index to the color table, respectively.

The growth of an internode requires an additional comment. Basically, productions p_3 and p_4 have the structure of the simple L-system:

$$\begin{aligned} \omega &: \quad a \\ p_1 &: \quad a \rightarrow b \\ p_2 &: \quad b \rightarrow ba \end{aligned}$$

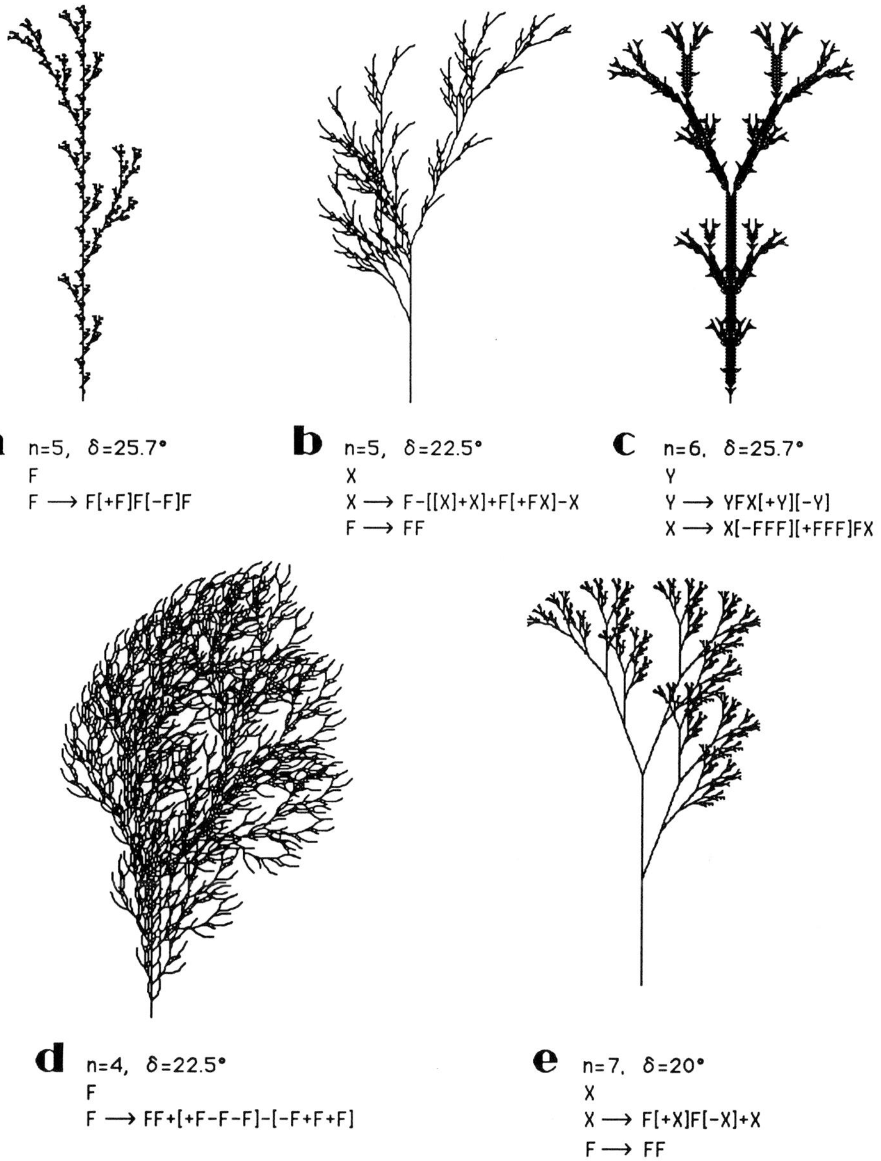

Figure 3.2: Examples of plant-like structures generated by bracketed OL-systems.

Figure 3.3: A three-dimensional *Fibonacci bush.*

It is known that this L-system generates strings of lengths coinciding with consecutive terms of the Fibonacci series $(1, 1, 2, 3, 5, 8, ...)$ [89]. Consequently, the lengths of internodes in the bush shown in Figure 3.3 conform to the same series.

Although various manifestations of the Fibonacci series occur frequently in nature [99], the particular growth rate characterizing the bush shown in Figure 3.3 was chosen purely for aesthetic reasons. The generated branching structure was supposed to look like a bush, but no attempt was made to model any existing species.

Plant with flowers

Another example of a three-dimensional plant generated by an OL-system is shown in Figure 3.4. The corresponding L-system is given below:

Figure 3.4: A plant generated by an L-system.

ω : **plant**
p_1 : **plant** $\rightarrow$ **internode** + [**plant** + **flower**] −− //
[−− **leaf**] **internode** [+ + **leaf**] −
[**plant flower**] + + **plant flower**
p_2 : **internode** $\rightarrow$ F **seg** [// & & **leaf**] [// ∧∧ **leaf**] F **seg**
p_3 : **seg** $\rightarrow$ **seg** F **seg**
p_4 : **leaf** $\rightarrow$ [' { $+f - ff - f+$ | $+f - ff - f$ }]
p_5 : **flower** $\rightarrow$ [& & & **pedicel** ' / **wedge** //// **wedge** ////
wedge //// **wedge** //// **wedge**]
p_6 : **pedicel** $\rightarrow$ FF
p_7 : **wedge** $\rightarrow$ [' ∧ F] [{ & & & & $-f + f$ | $-f + f$ }]

The angle increment δ is equal to 18°. This L-system can be described and analyzed in a way similar to the previous one. The exponential growth of the internodes (specified by production p_3) resulted in the "best-looking" plant in this case.

Tropism

A characteristic feature of turtle interpretation is that directions are specified relative to the current orientation. However, absolute directions play an important role in the development of plants. For example, branches may bend up towards the source of light or down due to gravity. These effects can be simulated by slightly rotating the turtle in the direction of a predefined ***tropism vector*** $\vec{T}$ after drawing each segment. In the two-dimensional case the orientation adjustment α is calculated from the formula $\alpha = e\vec{F} \times \vec{T}$, where e is a parameter capturing axis susceptibility to bending. This heuristic formula has a physical motivation; if $\vec{T}$ is interpreted as a force applied to the endpoint of segment $\vec{F}$ and $\vec{F}$ can rotate around its starting point, the torque is equal to $\vec{F} \times \vec{T}$. The effect of tropism on a branching structure is illustrated in Figure 3.5.

3.2 Developmental approach to plant modelling

3.2.1 Introduction

Problem statement

The examples presented in the previous section show the potential of using L-systems for plant modelling. They also illustrate one of the most striking features of the generative approach to modelling, called *data base amplification* [98]. This term refers to the generation of complex-looking objects from very concise descriptions — in our case, L-systems comprising a small number of productions. Although the L-systems under consideration are so concise, their construction is not a trivial task. While some L-systems which generate interesting plant-like structures can be obtained by trial-and-error,

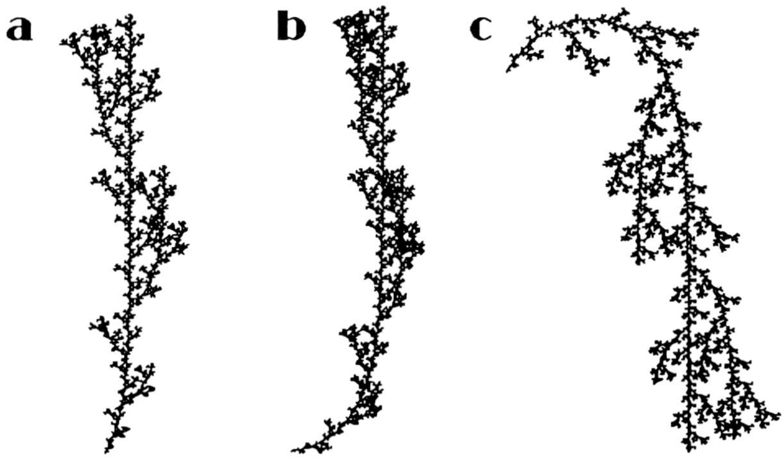

Figure 3.5: Modelling tropism. The tropism vector $\vec{T}$ points up. The coefficients e used to generate structures a–c satisfy the relation $e_b > e_a > 0 > e_c$.

it is advantageous to have a more systematic approach to the modelling of plants. The methodology presented in subsequent sections is based on the simulation of the development of real plants. Thus, in order to model a particular form, we attempt to capture the essence of the *developmental process* which leads to this form.

Growth and form

The view that growth and form are interrelated has a long tradition in biology. D'Arcy Thompson [102] traces its origins to the late seventeenth century, and comments:

> The rate of growth deserves to be studied as a necessary preliminary to the theoretical study of form, and organic form itself is found, mathematically speaking, to be a function of time... We might call the form of an organism an *event in space-time*, and not merely a *configuration in space.*

This concept is echoed in a sentence by Hallé, Oldeman and Tomlinson [35]:

> The idea of the form implicitly contains also the history of such a form.

Accordingly, the developmental approach to plant modelling has the following distinctive features:

- **Emphasis on the space-time relation between plant parts.** In many plants, various developmental stages can be observed at the same time. For example, some flowers may still be in the bud stage, others may be fully developed, and still others may have been transformed into fruits. If the developmental technique is consistently used down to the level of individual organs, such *phase effects* are reproduced in a natural way.

- **Inherent capability of growth simulation.** Since the entire developmental process is captured by the mathematical model, it can be used to generate biologically correct images of plants at different ages and to provide animated growth sequences.

Herbaceous plants

The formalism of L-systems is particularly suitable for the modelling and simulating the development of *herbaceous* or non-woody plants.

- Genetic factors play a predominant role in the development of herbaceous plants. In contrast, the form of woody plants is determined to a large extent by the environment, competition between trees and tree branches, and accidents [113]. These effects are unrelated to the control mechanisms considered in these notes.

- The development of woody plants is more complex than that of herbaceous plants because of secondary growth, which is responsible for the gradual increase of branch diameter with time.

- It is difficult to collect enough data to construct adequate developmental models of trees because of their long life cycle.

Control mechanisms in plants

In order to faithfully reenact plant development, we simulate natural control mechanisms. In biology, they are divided into two classes, called *lineage* and *interactive* mechanisms. The term *lineage* (or *cellular descent*) refers to the transfer of genetic information from an ancestor cell to its descendants. In contrast, *interaction* is a mechanism in which information is exchanged between coexisting neighbouring cells (for example, in the form of nutrients or hormones). Within the formalism of L-systems, lineage mechanisms are represented by context-free productions found in OL-systems, while the simulation of interaction requires the use of context-sensitive L-systems, known as bracketed *1L-systems* and *2L-systems*. We will first discuss developmental effects which can be modelled without interactions, then define the notion of context in bracketed L-systems and apply it to simulate development with interactions.

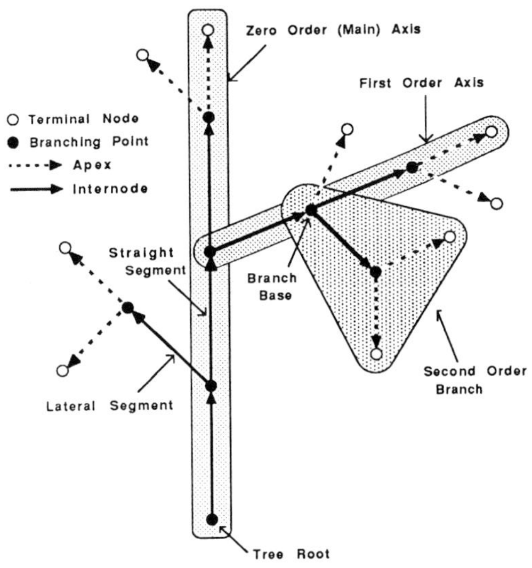

Figure 3.6: An axial tree.

3.2.2 Graph-theoretical vs. botanical trees

In the context of plant modelling, a branching structure or "tree" must be carefully defined to avoid ambiguity. To this end, we introduce the notion of an axial tree (Figure 3.6), which complements the graph-theoretic notion of a rooted tree [80] with the botanically motivated notion of branch axis.

A *rooted tree* has edges which are labelled and directed. The edge sequences form paths from a distinguished node called the *root* or the *base* to the *terminal nodes*. In the biological context, these edges are referred to as *branch segments*. A segment followed by at least one more segment in some path is called an *internode*. A terminal segment (with no succeeding edges) is called an *apex*.

An *axial tree* is a special type of rooted tree. At each of its nodes we distinguish at most one outgoing *straight* segment. All remaining edges are called *lateral* or *side* segments. A sequence of segments is called an *axis* if:

- the first segment in the sequence originates at the root of the tree or as a lateral segment at some node,
- each subsequent segment is a straight segment, and
- the last segment is not followed by any straight segment in the tree.

Together with all its descendants, an axis constitutes a *branch.* A branch is itself an axial (sub)tree.

Axes and branches are ordered. The axis originating at the root of the entire plant has order zero. An axis originating as a lateral segment of an n-order parent axis has order $n + 1$. The order of a branch is equal to the order of its lowest-order or *main* axis.

Axial trees are purely topological objects. The geometric connotation of such terms as straight segment, lateral segment and axis should be viewed at this point as an intuitive link between the graph-theoretic formalism and real plant structures.

3.3 Development without interactions

Schemata of L-systems

When discussing various types of plants, we will focus on the topological description of their structure and the temporal aspects of their development. Thus, the L-systems considered in the following sections should no longer be considered as "ready-to-use" data files. This departure from detailed listings is motivated by three factors:

- Abstraction from geometric details makes it easier to focus on the essential features inherent to a given type of development;
- The discussion of development becomes clearer if the symbols in the L-systems refer to the role particular segments play in the plant (e.g. an internode, an apex, a leaf) rather than to the details of turtle interpretation;
- Detailed L-systems corresponding to particular plant species tend to be much longer (and less legible) than their abstract "schemata".

Given an L-system scheme, a "detailed" L-system can be specified by including the symbols which describe the growth rates of internodes, the branching angles, and the structure of organs. Details on organ modelling are discussed in Chapter 4.

Terminology

We put particular emphasis on the modelling of compound flowering structures or *inflorescences.* As there is no commonly accepted terminology referring to inflorescence types, we chose to follow the terminology of Müller-Doblies [75], which in turn is based on extensive work by Troll [103]. Our presentation is organized by the control mechanisms which govern inflorescence development.

3.3.1 Racemes, or the phase beauty of sequential growth

The simplest possible flowering structures with multiple flowers are those with a single stem on which an indefinite number of flowers are produced sequentially. Inflorescences of this type are called *racemes.* Their development can be described by the following OL-system:

$$\begin{aligned} \omega &: \ A \\ p_1 &: \ A \to I[IF_0]A \\ p_2 &: \ F_i \to F_{i+1} \qquad i \geq 0 \end{aligned}$$

The edge symbol A denotes the apex of the main (zero-order) axis, I is an internode, and symbols F_i refer to subsequent stages of flower development. The indexed notation $F_i \to F_{i+1}$ stands for a (potentially infinite) set of productions $F_0 \to F_1, F_1 \to F_2, F_2 \to F_3, \ldots$. The developmental sequence begins as follows:

$$\begin{aligned} &A \\ &I[IF_0]A \\ &I[IF_1]I[IF_0]A \\ &I[IF_2]I[IF_1]I[IF_0]A \\ &I[IF_3]I[IF_2]I[IF_1]I[IF_0]A \cdots \end{aligned}$$

Lily-of-the-valley

It can be seen that at each developmental stage the inflorescence contains a sequence of flowers of different ages. The flowers newly created by the apex are delayed in their development with respect to the older ones situated at the stem base. Graphically, this effect is illustrated by the model of a lily-of-the-valley shown in Figure 3.7. Its inflorescence was generated by the above L-system complemented with attribute symbols to control geometry of internodes, branching angles, and organ structure. The following quotation from d'Arcy Thompson [102] applies:

> A flowering spray of lily-of-the-valley exemplifies a growth-gradient, after a simple fashion of its own. Along the stalk the growth-rate falls away; the florets are of descending age, from flower to bud; their graded differences of age lead to an exquisite gradation of size and form; the time-interval between one and another, or the "space-time relation" between them all, gives a peculiar quality — we may call it phase-beauty — to the whole.

Fern

This "phase-beauty" can also be observed in other structures. For example, consider the fern-like plant shown in Figure 3.8. In this case, nine zero-order branches grow subapically and produce new first-order branches, which also

Figure 3.7: Lily-of-the-valley.

Figure 3.8: Fern.

grow subapically and produce leaves. The number of leaves carried on first-order branches, the length of internodes and the leaf size increase with time. These processes are described by the following L-system:

$$\begin{array}{lll} \omega: & [A][A][A][A][A][A][A][A][A] & \\ p_1: & A \rightarrow I_0[B]A & \\ p_2: & B \rightarrow I_0[L_0][L_0]B & \\ p_3: & I_i \rightarrow I_{i+1} & i \geq 0 \\ p_4: & L_i \rightarrow L_{i+1} & i \geq 0 \end{array}$$

A and B denote apices of zero-order and first-order axes, $I_0, I_1, I_2, \ldots$ denote the internodes, and $L_0, L_1, L_2, \ldots$ denote the subsequent stages of leaf development.

3.3.2 Cymose inflorescences, or the use of delays

In racemes the apex of the main axis produces lateral branches and continues to grow. In contrast, the apex of the main axis in *cymes* turns into a flower shortly after a few lateral branches have been initiated. Their apices turn into flowers as well, and second-order branches take over. In time, branches of higher and higher order are produced. Thus, the basic structure of a cymose inflorescence is captured in the production:

$$A \rightarrow I[A][A]IF$$

A denotes an apex, I an internode, and F a flower. Note that according to this description, the two branches are identical and grow in concert. In reality, this need not be the case, as one lateral branch may start growing before the other. This effect can be modelled by assuming that apices undergo a sequence of state changes which delay their further growth until a particular state is reached. For example, consider the following L-system which describes the development of the rose campion (*Lychnis coronaria*) as analyzed by Robinson [84]: *Lychnis*

$$\begin{array}{lll} \omega: & A_7 & \\ p_1: & A_7 \rightarrow I[A_0][A_4]IF_0 & \\ p_2: & A_i \rightarrow A_{i+1} & 0 \leq i < 7 \\ p_3: & F_i \rightarrow F_{i+1} & i \geq 0 \end{array}$$

Production p_1 shows that at their creation time, the lateral apices have different states A_0 and A_4. Consequently, the first apex requires eight derivation steps to produce a flower and new branches, while the second requires only

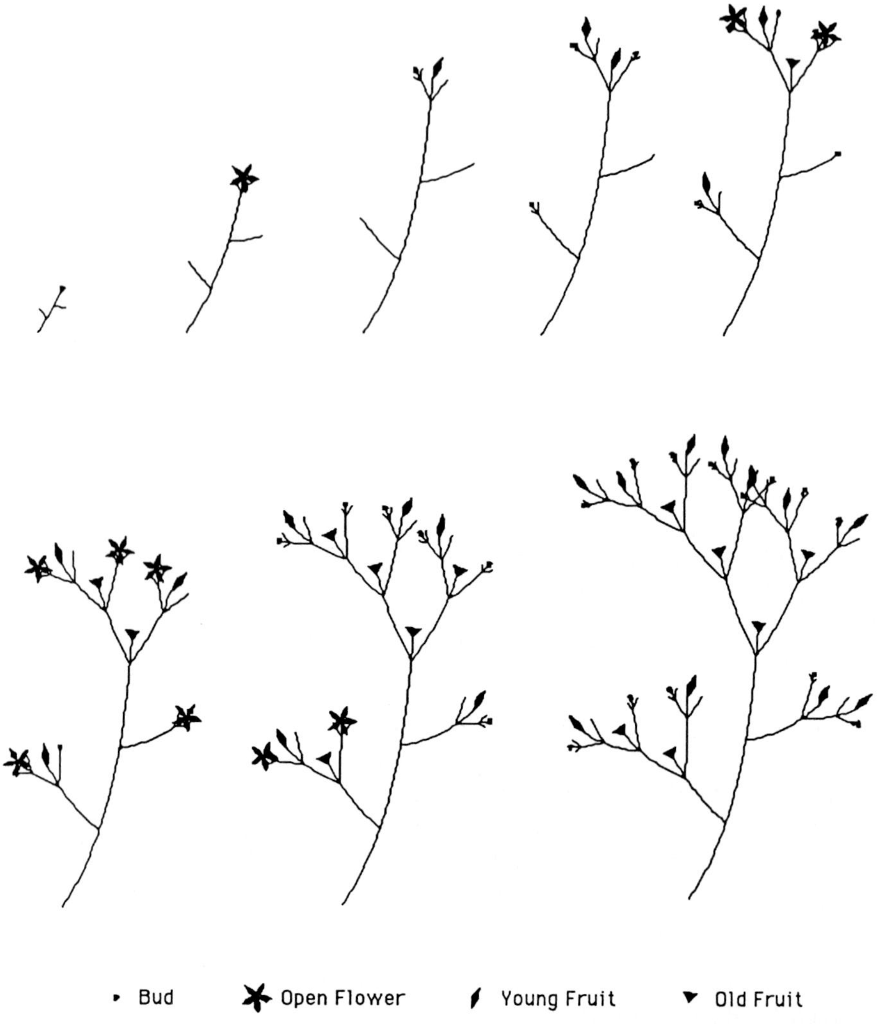

The inflorescence is composed of a central and two lateral sub-inflorescences. Every third derivation step is shown. Stages of flower development: F_0, F_1, F_2: bud, F_3, F_4: open flower, $F_5 - F_9$: young fruit, $F_{10}, F_{11}, \ldots$: old fruit.

Figure 3.9: Development of *Lychnis coronaria*.

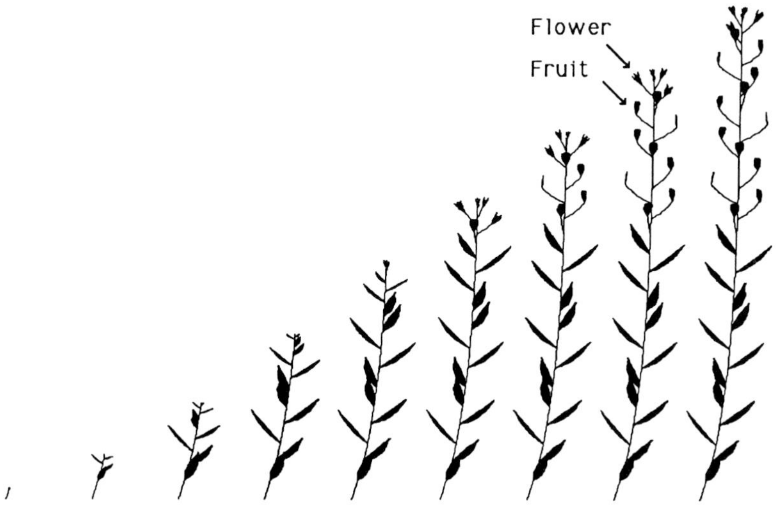

Every fourth derivation step is shown. Stages of flower development: $F_0 - F_4$: open flower. $F_5, F_6, \ldots$: fruit.

Figure 3.10: Development of *Capsella bursa-pastoris*.

four steps. Concurrently, each flower undergoes a sequence of changes, progressing from the bud stage to an open flower to a fruit. This developmental sequence is illustrated in Figure 3.9. From the biological perspective it is interesting to note that at different developmental stages there are a number of open flowers which have a relatively uniform distribution over the entire plant structure. This is advantageous to the plant as it increases the time span over which seeds will be produced.

3.3.3 Racemes with leaves, or modelling qualitative changes of the developmental process

The developmental sequences considered so far are homogeneous in the sense that the same structure is produced repeatedly at fixed time intervals. However, in many cases a qualitative change in the nature of development can be observed at some point in time. For example, consider shepherd's purse (*Capsella bursa-pastoris*) shown in Figure 3.10. In principle, its developmen- *Capsella* tal pattern can be described as follows:

$$
\begin{array}{lll}
\omega: & A & \\
p_1: & A \rightarrow I[L_0]A & \\
p_2: & A \rightarrow I[L_0]B & \\
p_3: & B \rightarrow I[IF_0]B & \\
p_4: & L_i \rightarrow L_{i+1} & i \geq 0 \\
p_5: & F_i \rightarrow F_{i+1} & i \geq 0
\end{array}
$$

The initial vegetative growth, represented by production p_1, describes the creation of successive internodes and leaves by apex A. At some point in time, production p_2 changes the apex from the vegetative state A to the flowering state B. From then on, flowers are produced instead of leaves, forming a raceme structure as discussed in Section 3.3.1. The moment in which this change (or developmental switch) occurs is not specified: the L-system is a nondeterministic one. Thus, for modelling purposes it must be complemented with an additional control mechanism which will determine the switch time. Below we outline three possible mechanisms. Each of them is biologically motivated, and corresponds to a different class of L-systems.

Delay mechanism

As in the case of cymose inflorescences (Section 3.3.2), the apex undergoes a series of state changes which delay the switch until a particular state is reached. This mechanism is outlined below:

$$
\begin{array}{lll}
\omega: & A_0 & \\
p_1: & A_i \rightarrow I[L_0]A_{i+1} & 0 \leq i < n \\
p_2: & A_n \rightarrow I[L_0]B & \\
p_3 - p_5: & \textit{as before} &
\end{array}
$$

According to this model, the apex *counts* the leaves it produces. While it may seem strange that a plant counts, it is known that some plant species produce a fixed number of leaves before they start flowering. Counting is achieved by the monotonic increase or decrease of the concentration of certain cell components.

Stochastic mechanism

Another method for implementing the change is to use a stochastic mechanism. In this case the vegetative apex has a probability π_1 of staying in the vegetative state and π_2 of transforming itself into a flowering apex:

$$\begin{array}{ll} \omega : & A \\ p_1 : & A \xrightarrow{\pi_1} I[L_0]A \\ p_2 : & A \xrightarrow{\pi_2} I[L_0]B \\ p_3 - p_5 : & \textit{as before} \end{array}$$

For a formal definition of stochastic L-systems see [21,112].

Environmental mechanism

Many plants change from the vegetative to the flowering state in response to an environmental factor (such as the number of daylight hours or temperature). We can model this by using one set of productions (called a *table*) for some number of derivation steps before replacing it by another table:

Table 1			**Table 2**		
$\omega :$	A		$p_1 :$	$A \rightarrow I[L_0]B$	
$p_1 :$	$A \rightarrow I[L_0]A$		$p_2 :$	$B \rightarrow I[IF_0]B$	
$p_2 :$	$L_i \rightarrow L_{i+1}$	$i \geq 0$	$p_3 :$	$L_i \rightarrow L_{i+1}$	$i \geq 0$
			$p_4 :$	$F_i \rightarrow F_{i+1}$	$i \geq 0$

The concept of table L-systems (*TOL-systems*) is formalized in [37,88].

The developmental switch from the vegetative to the flowering state is not the only qualitative change which can occur in a plant. Another possibility is the transformation of an apex from producing lateral flowers to producing a terminal flower which stops the axis development. This switch can also be produced using the methods described above.

3.3.4 Composition of interactionless inflorescences

In Section 3.3.1 we considered simple racemes (also called *monobotryoid* inflorescences). They are characterized by a single axis on which flowers are borne. In many plants compound racemes also occur. In the *dibotryoid* case the main axis carries entire racemes on the first order axes. This composition can be recursively extended to higher orders (*tribotryoids*, etc.). In general, the L-system for a compound raceme of order n has the following productions:

$$\begin{array}{lll} p_1 : & A_i \rightarrow I[A_{i+1}]A_i & 0 \leq i \leq n-1 \\ p_2 : & A_n \rightarrow I[IF]A_n & \end{array}$$

The above L-system does not produce flowers on axes of order less then n. However, some plants develop racemes at the end of each axis. This can be modelled by introducing additional productions of the form:

$$p'_1: \; A_i \rightarrow I[IF]A_n \quad 0 \leq i \leq n-1$$

The mechanism for switching from productions p_1 to productions p'_1 can be the same as those discussed in Section 3.3.3.

Similar rules for inflorescence composition apply to cymes. In fact, composition was used to produce two lower branches of *Lychnis coronaria* shown in Figure 3.9.

3.4 Context-sensitive L-systems

Productions in OL-systems are context-free; i.e. applicable regardless of the context in which the predecessor appears. This type of production is sufficient to model cellular descent, that is, information transfer from the parent cell to its descendants. However, a context-sensitive extension of L-systems is necessary to model information exchange between neighbouring cells (cellular interactions). Various possible extensions have been proposed and thoroughly studied in the past [89,37,61]. Some of these are listed below. *2L-systems* use productions of the form $a_l < a > a_r \rightarrow \chi$, where the letter a (called the *strict predecessor*) can produce word χ if and only if a is preceded by letter a_l and followed by a_r. Thus, letters a_l and a_r form the left and the right *context* of a in this production. Productions in *1L-systems* have one-sided context only; consequently, they are either of the form $a_l < a \rightarrow \chi$ or $a > a_r \rightarrow \chi$. OL-systems, 1L-systems and 2L-systems belong to a wider class of *IL-systems*, also called *(k,l)-systems*. In a (k,l)-system, the left context is a word of length k and the right context is a word of length l.

In order to keep specifications of L-systems short, we slightly modify the usual notion of IL-systems by allowing productions with different context lengths to coexist within a single system. Furthermore, we assume that context-sensitive productions have precedence over context-free productions with the same strict predecessor. Consequently, if a context-free and a context-sensitive production both apply to a given letter, the context-sensitive one should be selected. If no production applies, this letter is replaced by itself as previously assumed for OL-systems.

Context in strings

As an example of cellular interaction, consider the diffusion of a hormone along a filament. If a denotes a cell with hormone concentration below a threshold level and b is a cell with concentration exceeding this level, the diffusion process can be described by the following 1L-system:

$$\begin{aligned} \omega &: \ baaaaaaa \\ p &: \ b < a \rightarrow b \end{aligned}$$

The first few words generated by this L-system are given below:

baaaaaaaa
bbaaaaaaa
bbbaaaaaa
bbbbaaaaa
bbbbbaaaa
...

Thus, the hormone propagates throughout the filament, starting from its left end.

Context in trees

Context-sensitive bracketed L-systems are comparatively more complex than L-systems without brackets, because symbols representing adjacent tree segments can be separated by an arbitrarily large number of other symbols in the bracketed string representation. Consequently, special rules for context searching are needed. We will consider a restricted case where daughter branches do not belong to the context of the mother branch. This approach corresponds to the original definition of bracketed L-systems with interactions [54]. For example, in the string:

$$ABCD[EF][G[HI[JKL]M]NOPQ]$$

D is the left context of G, and N is the right context of G.

L-systems of Hogeweg, Hesper and Smith

In 1974 Hogeweg and Hesper published the results of an exhaustive study of 3584 patterns generated by a class of bracketed 2L-systems defined over the alphabet {0,1} [38]. Some of these patterns had plant-like shapes. Subsequently, Smith significantly improved the quality of the generated images using state-of-the-art computer imagery techniques [94,98]. Examples of structures obtained using the L-systems of Hogeweg and Hesper are shown in Figure 3.11. The only modification required was to incorporate the geometric information needed to control the turtle. The geometric symbols are ignored while context matching.

The complete data files used to produce the plant-like structures shown in Figure 3.11 are listed below.

a
```
derivation length: 30
angle factor: 16
scale factor: 100
axiom: F1F1F1
ignore: +-F
0 < 0 > 0 --> 0
0 < 0 > 1 --> 1[+F1F1]
0 < 1 > 0 --> 1
0 < 1 > 1 --> 1
1 < 0 > 0 --> 0
1 < 0 > 1 --> 1F1
1 < 1 > 0 --> 0
1 < 1 > 1 --> 0
* < + > * --> -
* < - > * --> +
```

b
```
derivation length: 30
angle factor: 16
scale factor: 100
axiom: F1F1F1
ignore: +-F
0 < 0 > 0 --> 1
0 < 0 > 1 --> 1[-F1F1]
0 < 1 > 0 --> 1
0 < 1 > 1 --> 1
1 < 0 > 0 --> 0
1 < 0 > 1 --> 1F1
1 < 1 > 0 --> 1
1 < 1 > 1 --> 0
* < - > * --> +
* < + > * --> -
```

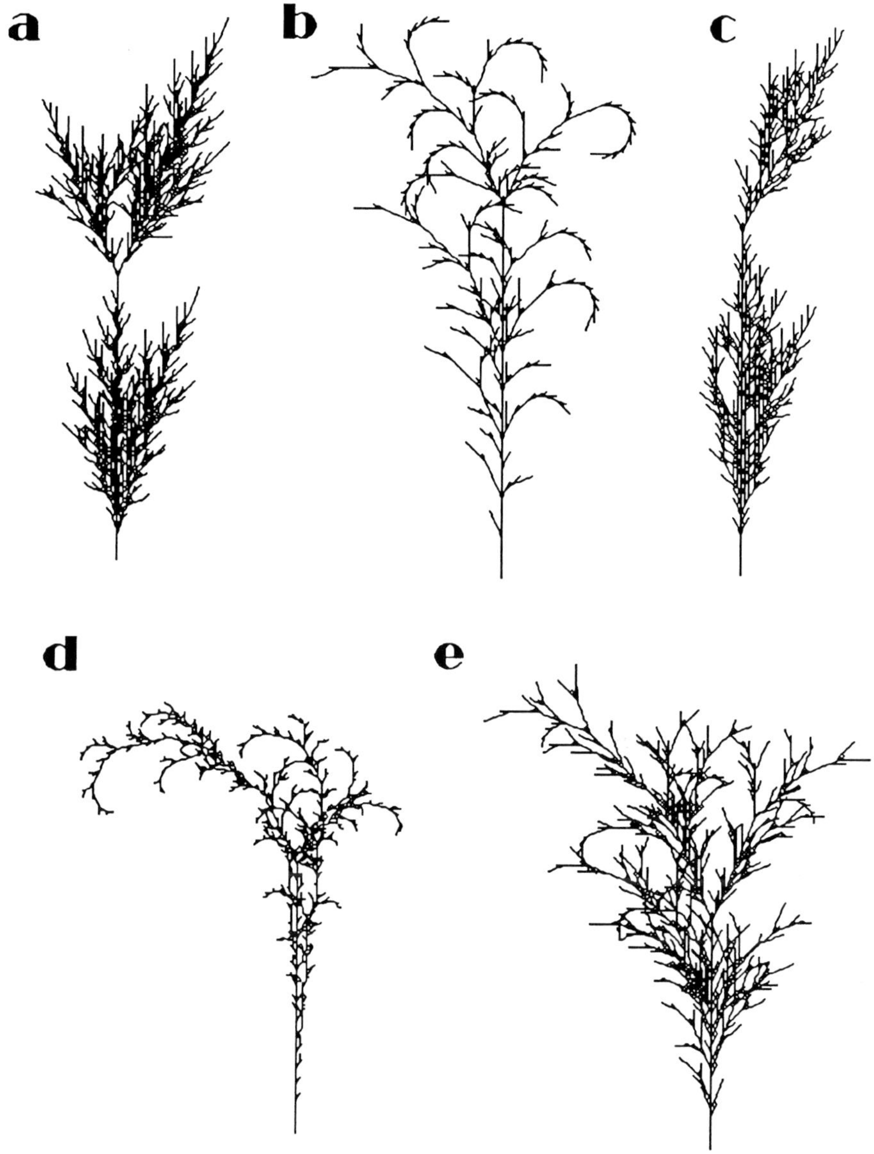

Figure 3.11: Examples of branching structures generated using the L-systems of Hogeweg and Hesper.

c

```
derivation length:  26
angle factor:  14
scale factor:  100
axiom:  F1F1F1
ignore:  +-F
0 < 0 > 0 --> 0
0 < 0 > 1 --> 1
0 < 1 > 0 --> 0
0 < 1 > 1 --> 1[+F1F1]
1 < 0 > 0 --> 0
1 < 0 > 1 --> 1F1
1 < 1 > 0 --> 0
1 < 1 > 1 --> 0
* < - > * --> +
* < + > * --> -
```

d

```
derivation length:  24
angle factor:  14
scale factor:  100
axiom:  F0F1F1
ignore:  +-F
0 < 0 > 0 --> 1
0 < 0 > 1 --> 0
0 < 1 > 0 --> 0
0 < 1 > 1 --> 1F1
1 < 0 > 0 --> 1
1 < 0 > 1 --> 1[+F1F1]
1 < 1 > 0 --> 1
1 < 1 > 1 --> 0
* < + > * --> -
* < - > * --> +
```

e

```
derivation length:  30
angle factor:  16
scale factor:  100
axiom:  F1F1F1
ignore:  +-F
0 < 0 > 0 --> 0
0 < 0 > 1 --> 1[-F1F1]
0 < 1 > 0 --> 1
0 < 1 > 1 --> 1
1 < 0 > 0 --> 0
1 < 0 > 1 --> 1F1
1 < 1 > 0 --> 1
1 < 1 > 1 --> 0
* < + > * --> -
* < - > * --> +
```

The exhaustive search of a particular class of L-systems made it possible to find several which produce fairly realistic images, but did not offer a methodology for *constructing* context-sensitive L-systems for modelling given plant structures. In order to develop such a methodology, we must analyze the role of interaction in plant development.

3.5 Development with interactions

3.5.1 Signals in plants

Need for interaction

Even in the presence of delays, lineage mechanisms (corresponding to context free productions) reflect the sequential creation of branches, flowers and leaves by the subapical growth process. Consequently, organs near the plant roots develop earlier and more extensively than those situated near the axis ends. Such development results in *basitonic* plant structures (heavily developed near the base) with *acropetal* flowering sequences (the zone of blooming flowers progresses upwards along each branch). However, nature also creates *acrotonic* structures (heavily developed near the apex) and *basipetal* flowering sequences (progressing downwards). These structures and developmental patterns cannot be viewed as a simple consequence of subapical growth; for example, basipetal flowering sequences progress in the direction which is precisely opposite to that of plant growth. An intuitively straightforward and biologically well founded explanation of the described phenomena can be given in terms of signals which propagate through the plant and control the timing of developmental processes.

Signal propagation

Within the formalism of bracketed L-systems, the left context can be used to simulate control signals which propagate acropetally, i.e. from the root or the basal leaves towards the apices of the modelled plant, while the right context represents signals which propagate basipetally, i.e. from the apices towards the root. For example, the following 1L-system simulates propagation of an acropetal signal in a branching structure which does not grow.

$$\begin{aligned} \omega &: \ J[I]I[I]I[I]I \\ p &: \ J < I \rightarrow J \end{aligned}$$

Symbol J represents a segment already reached by the signal, while I represents a segment which has not yet been reached. Images representing consecutive stages of signal propagation (corresponding to consecutive words generated by the L-system under consideration) are shown in Figure 3.12.

The propagation of a basipetal signal can be simulated in a similar way (Figure 3.13):

$$\begin{aligned} \omega &: \ I[I]I[I]I[I]J \\ p &: \ I > J \rightarrow J \end{aligned}$$

Below we consider two developmental models with signals. The first model employs a single acropetal signal, while the second one uses both acropetal and basipetal signals. In contrast to the previous examples we assume that control signals propagate in a structure which is growing.

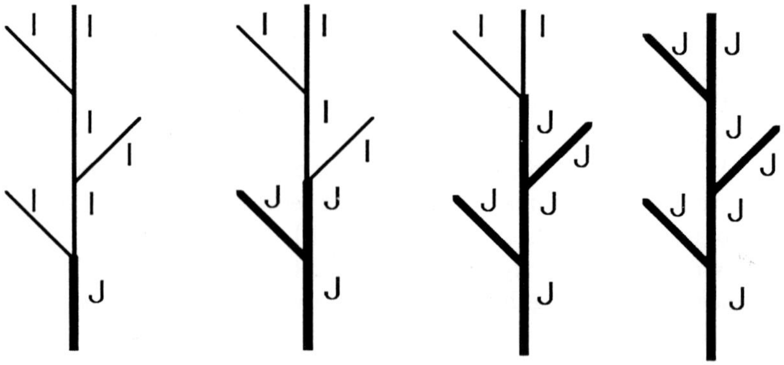

Figure 3.12: Acropetal signal propagation.

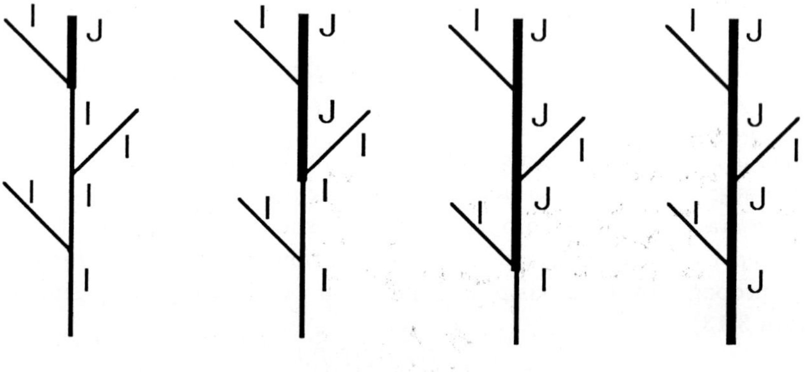

Figure 3.13: Basipetal signal propagation.

3.5.2 Developmental model with a single acropetal signal

Let us assume that the switch from the vegetative to the flowering condition is caused by a flower-inducing signal (representing the hormone *florigen*), which is transported from the basal leaves towards branch apices. In this case, the overall phase effect results from an interplay between growth and control signal propagation [58,39]. Assuming that only the first-order lateral branches are present, the development can be described by the following 1L-system:

$$
\begin{array}{lll}
\omega: & D_0A_0 & \\
p_1: & A_i \rightarrow A_{i+1} & 0 \le i < m-1 \\
p_2: & A_{m-1} \rightarrow I[B_0]A_0 & \\
p_3: & B_i \rightarrow B_{i+1} & 0 \le i < n-1 \\
p_4: & B_{n-1} \rightarrow J[L]B_0 & \\
p_5: & D_i \rightarrow D_{i+1} & 0 \le i < d \\
p_6: & D_d \rightarrow S_0 & \\
p_7: & S_i \rightarrow S_{i+1} & 0 \le i < \max\{u,v\}-1 \\
p_8: & S_z \rightarrow \epsilon & z = \max\{u,v\}-1 \\
p_9: & S_{u-1} < I \rightarrow IS_0 & \\
p_{10}: & S_{v-1} < J \rightarrow JS_0 & \\
p_{11}: & S_0 < A_i \rightarrow F_0 & 0 \le i \le m-1 \\
p_{12}: & S_0 < B_i \rightarrow F_0 & 0 \le i \le n-1 \\
p_{13}: & F_i \rightarrow F_{i+1} & i \ge 0
\end{array}
$$

This L-system operates as follows. The apex A produces segments I of the main axis and creates the lateral apices (p_1, p_2). The time between the production of two consecutive segments, called the *plastochron* of the main axis, is equal to m units (derivation steps). In a similar way, the first-order apices B produce segments J of the lateral axes and leaves L with plastochron n (p_3, p_4). After a delay of d time units a signal S is sent from the tree base towards the apices (p_6). This signal is transported along the main axis with a delay of u time units per internode I (p_7, p_9), and along the first-order axes with a delay of v units per internode J (p_7, p_{10}). Production p_8 removes the signal from a node after it has been transported further along the structure (ϵ stands for the empty string). When the signal reaches an apex (either A or B), the apex is transformed into a terminal flower F (p_{11}, p_{12}) which undergoes the usual sequence of states (p_{13}).

Model analysis

In order to analyze the plant structure and flowering sequence resulting from the above development, let T_k denote the time at which the apex of the k-th first-order axis is transformed into a flower, and l_k denote the length of this axis (expressed as the number of internodes) at the transformation time. Since it takes km time units to produce k internodes along the main axis and $l_k n$ time units to produce l_k internodes on the first-order axis, we obtain:

$$T_k = km + l_k n$$

On the other hand, the transformation occurs when the signal S reaches the apex. The signal is sent d time units after the development starts. It uses ku time units to travel through k zero-order internodes and $l_k v$ time units to travel through l_k first-order internodes:

$$T_k = d + ku + l_k v$$

Solving the above system of equations for l_k and T_k (and ignoring for simplicity some inaccuracy due to the fact that this system does not guarantee integer solutions), we obtain:

$$T_k = k\frac{un - vm}{n - v} + d\frac{n}{n - v}$$

$$l_k = -k\frac{m - u}{n - v} + \frac{d}{n - v}$$

In order to analyze the above solutions, let us first notice that the signal transportation delay v must be less than the plastochron of the first-order axes n. If this were not the case the signal would never reach the apices. Under this assumption, the sign of the expression $\Delta = un - vm$ determines the flowering sequence, which is acropetal for $\Delta > 0$ and basipetal for $\Delta < 0$ (Figure 3.14). If $\Delta = 0$, all flowers occur simultaneously. The sign of the expression $m - u$ determines whether the plant has a basitonic $(m - u < 0)$ or acrotonic $(m - u > 0)$ structure.

3.5.3 Developmental model with several signals

The development of some inflorescences is controlled by several signals, which may propagate with different delays and trigger each other. The use of more than one signal is instrumental in the modelling of a large class of inflorescences (found, for instance, in the family Compositae) characterized by terminal flowers on all apices, indefinite order of branching, and a basipetal flowering sequence. Figure 3.15 illustrates this type of development with an model of wall lettuce (*Mycelis muralis*). First, the main axis is formed in a process of subapical growth which produces subsequent internodes and lateral apices (a). At this stage further development of lateral branches is suppressed by *apical dominance*, that is, the inhibiting effect of an active apex exercised on the lateral branches of the same axis. Physiologically, the

Mycelis

An acropetal sequence: $m = 2, n = 3, u = v = 1, \Delta = 0.5$; derivation lengths: $15 - 19 - 21 - 23 - 26$.

A basipetal sequence: $m = 2, n = 5, u = 1, v = 3, \Delta = -0.5$; derivation lengths: $10 - 16 - 23 - 25 - 28 - 31$.

Figure 3.14: Flowering sequences generated by the model with an acropetal signal. Stages of flower development: F_0: bud (small circle), F_1, F_2, F_3: open flower, $F_4, F_5, \ldots$: fruit (large circle).

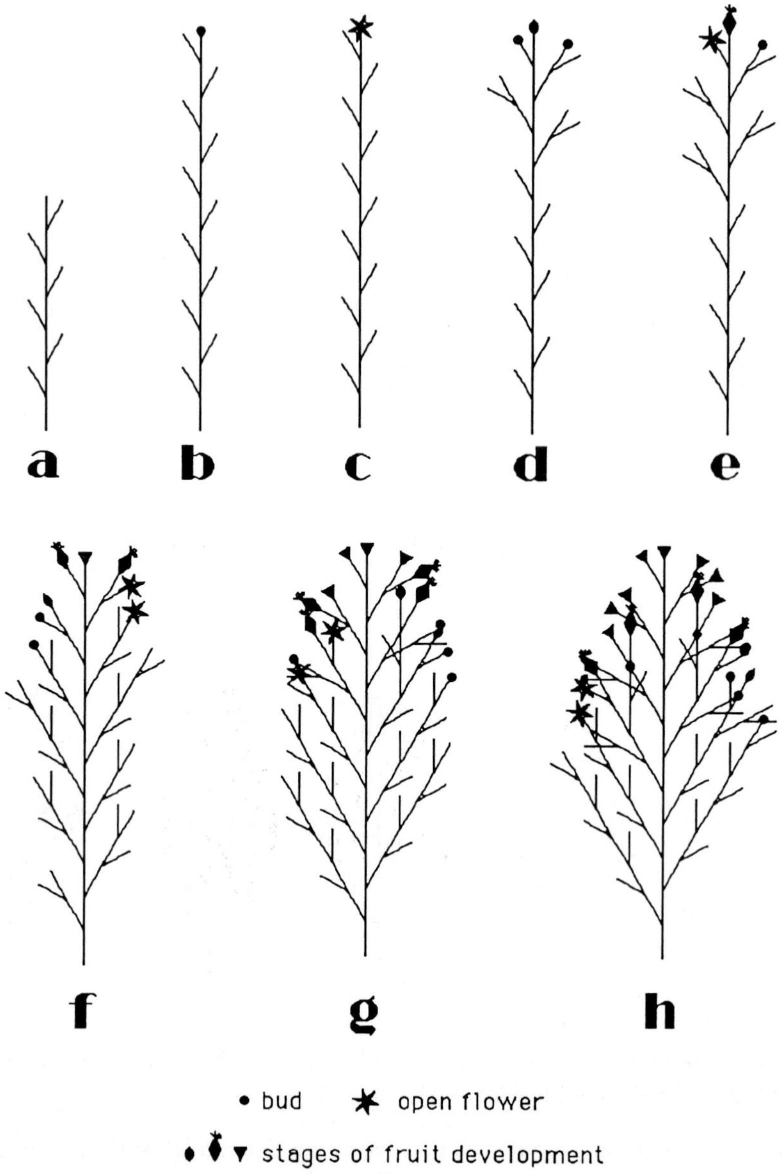

Figure 3.15: Developmental sequence of *Mycelis muralis*.

apical dominance is related to hormones called *auxins* present in the plant as long as the apex is active and gradually neutralized after flower initiation. At some moment, a flowering signal S_1 is sent from the bottom of the inflorescence along the main axis. When this signal reaches the apex, the terminal flower is initiated **(b,c)** and a basipetal signal S_2 lifting the apical dominance is sent down the main axis. This enables successive first-order axes to grow, starting from the topmost one **(d)**. After a delay, a secondary basipetal signal S_3 is sent from the apex of the main axis. Its effect is to send the flowering signal S_1 along subsequent first-order axes as they are encountered on the way down and, consequently, induce flowering at their apices **(e)**. This entire process repeats recursively for each axis: its apex is transformed into a flower, the apical dominance is lifted enabling lateral axes of the next order to grow, and the secondary basipetal signal is sent to induce the flowering signal S_1 in these lateral axes **(f,g,h)**. The resulting structures depend heavily on the values of plastochrons, delays, and signal propagation times. In the example under consideration, signal S_2 travels faster than S_3. Consequently, the time interval between the arrival of signals S_2 and S_3 increases while moving down the plant, potentially allowing the lower axes to grow longer than the upper ones. On the other hand, the lower branches start developing later, so they may not be fully developed at the time of observation. As a result of these opposite tendencies, the plant is developed most extensively in its middle parts. For a detailed biological analysis of the above process see [39].

3.6 Adding variation to models

All plants generated by the same deterministic L-system are identical. An attempt to combine them in the same picture would produce a striking, artificial regularity. In order to prevent this effect, it is necessary to introduce specimen-to-specimen variations which will preserve the general aspects of a plant but will modify its details.

Variation can be achieved by randomizing the turtle interpretation, the L-system, or both. Randomization of the interpretation alone has a limited effect. While the geometric aspects of a plant — such as the stem lengths and branching angles — are modified, the underlying topology remains unchanged. In contrast, the stochastic application of productions mentioned in Section 3.3.3 may affect both the topology and the geometry of the plant. We will limit our discussion to the context free case.

A *stochastic OL-system* is an ordered quadruplet $G_\pi = \langle V, \omega, P, \pi \rangle$. The

Stochastic L-system

alphabet V, the axiom ω and the set of productions P are defined as in an OL-system (Section1.2). Function $\pi : P \rightarrow (0, 1]$, called the ***probability distribution***, maps the set of productions into the set of ***production probabilities***. It is assumed that for any letter $a \in V$, the sum of probabilities of all productions with the predecessor a is equal to 1.

The above definition of a stochastic OL-system is similar to that of Yokomori [112] and Eichhorst and Savitch [21].

Stochastic derivation

We will call the derivation $\mu \Rightarrow \nu$ a *stochastic derivation* in G_π if for each *occurrence* of the letter a in the word μ the probability of applying production p with the predecessor a is equal to $\pi(p)$.

According to the above definition, different productions with the same predecessor can be applied to various occurrences of the same letter in one derivation step.

A simple example of a stochastic L-system is given below.

$$
\begin{aligned}
\omega &: F \\
p_1 &: F \xrightarrow{0.33} F[+F]F[-F]F \\
p_2 &: F \xrightarrow{0.33} F[+F]F \\
p_3 &: F \xrightarrow{0.34} F[-F]F
\end{aligned}
$$

The production probabilities are listed above the derivation symbol $\longrightarrow$. Each production can be selected with approximately the same probability of 1/3. Examples of branching structures generated by this L-system with derivations of length 5 are shown in Figure 3.16. Note that these structures look like different specimens of the same (albeit fictitious) plant species.

Flower field

A more complex example is shown in Figure 3.17. The *flower field* consists of four rows and four columns of plants. All plants are generated by a stochastic modification of the L-system used to generate Figure 3.4. The essence of this modification is to replace the original production p_3 by the following three productions:

$$
\begin{aligned}
p_3' &: \ \mathbf{seg} \xrightarrow{0.33} \mathbf{seg}\ [\ //\ \&\ \&\ \mathbf{leaf}\ ]\ [//\ \wedge\wedge\ \mathbf{leaf}\ ]\ F\ \mathbf{seg} \\
p_3'' &: \ \mathbf{seg} \xrightarrow{0.33} \mathbf{seg}\ F\ \mathbf{seg} \\
p_3''' &: \ \mathbf{seg} \xrightarrow{0.34} \mathbf{seg}
\end{aligned}
$$

Thus, in any step of the derivation, the stem segment **seg** may grow and produce new leaves (production p_3'), grow without producing new leaves (production p_3''), or not grow at all (production p_3'''). All three events occur

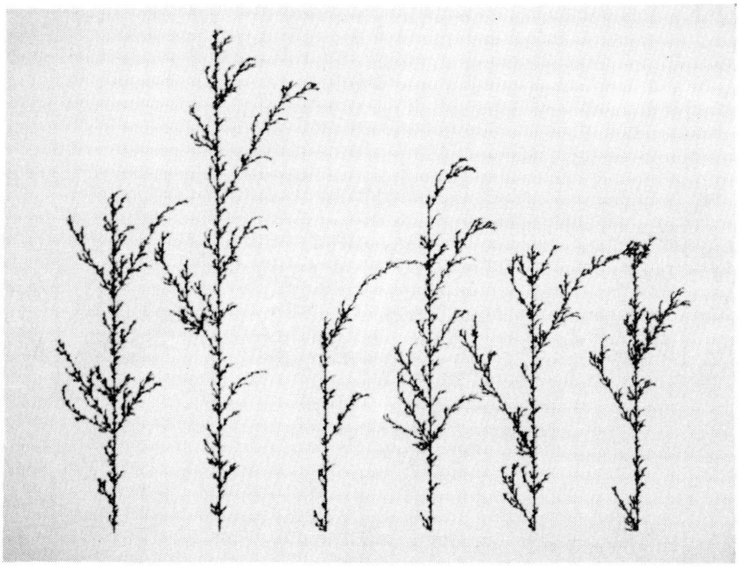

Figure 3.16: Sample stochastic L-system branching structures.

Figure 3.17: A flower field.

Figure 3.18: Shoots of the Japanese cypress.

with approximately the same probability. The resulting field appears to consist of various specimens of the same plant species.

Japanese cypress

While the stochastic L-systems discussed above were not intended to represent any existing plants, there are also stochastic L-systems developed by biologists to model particular plant species. Specifically, Nishida [78] proposed a stochastic L-system to model shoots of the Japanese cypress. The examples shown in Figure 3.18 were obtained using a slightly modified version of his model (productions with very small probabilities were eliminated).

Chapter 4

Models of plant organs

So far we have discussed the modelling of "skeletal" trees with branches consisting of mathematical lines. In this section we extend the model to include surfaces and volumes.

4.1 Bicubic surfaces

Conceptually, the simplest approach to organ modelling is to incorporate predefined surfaces into a tree structure. The standard method for describing arbitrary surfaces is based on bicubic patches [24]. Surface shapes are specified interactively, using a graphical patch editor. The alphabet of the L-system is extended to include symbols representing different patch shapes. When the turtle encounters a patch symbol during string interpretation, the corresponding patch is incorporated into the model with its position and orientation determined by the current state of the turtle. String symbols can also be used to control other features of the incorporated surface, such as its color or overall size. Thus, the L-system controls the appearance of the surfaces and their distribution in space, while the surface shapes are defined outside the conceptual framework of L-systems.

A simple example of a "generic plant" modelled using bicubic surfaces is shown in Figure 4.1. This plant structure is described by the following L-system:

$$
\begin{array}{lll}
\omega : & \textbf{apex} & \\
p_1 : & \textbf{apex} & \rightarrow \textbf{internode} \; [\, \textbf{leaf} \,] \; \textbf{apex} \\
p_2 : & \textbf{apex} & \rightarrow \textbf{internode bud} \\
p_3 : & \textbf{bud} & \rightarrow [\, \textbf{petal1} \,] \, [\, \textbf{petal2} \,] \, [\, \textbf{petal1} \,] \\
 & & \quad\; [\, \textbf{petal2} \,] \, [\, \textbf{petal1} \,] \, [\, \textbf{petal2} \,]
\end{array}
$$

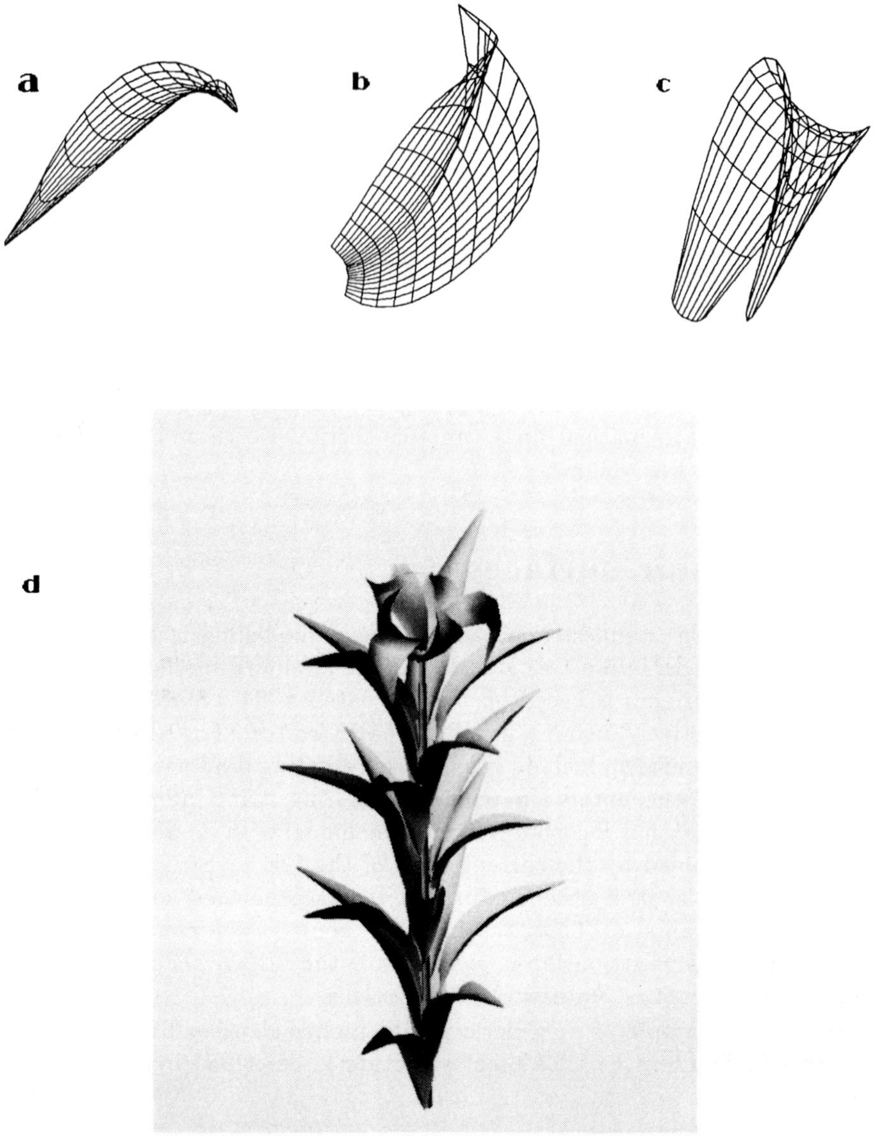

Figure 4.1: Construction of a plant. (a) Patch representing a leaf. (b,c) Patches representing petals. (d) The resulting plant.

Identifiers **leaf**, **petal1** and **petal2** denote patches incorporated in the model.

The simplicity of the above example makes it suitable for use in explaining the idea of incorporating predefined surfaces in a plant model. On the other hand, a model which consists of a handful of patches can easily be defined without any generative mechanism at all! However, patches can also be incorporated in much more complicated structures. For example, Figure 4.2 explains the structure of a lilac twig. The final image is shown in Fig. 4.3.

The topology and geometry of the model are based on observations of "real" lilac twigs. The process of inferring data from observations was not a trivial one. While the overall topology and the branching angles are constant within a given plant species, the lengths of internodes are known to vary substantially from one specimen to another. It is therefore difficult to capture the essential geometrical relationships characterizing a given structure from data affected by the "noise" of apparently random perturbations.

4.2 Developmental surface models

Patches make it easy to manipulate and modify surface shapes interactively, but are incompatible with the developmental approach to modelling since they do not "grow". Consequently, they are not suitable for use in models in which phase effects between organs play an important role.

In order to fully simulate plant development, it is necessary to provide a mechanism for changing the size and shape of surfaces in time. We can achieve this in a simple way by defining polygonal surface boundaries using an L-system and filling the resulting polygons. An example is presented below:

$$\begin{aligned} \omega &: \ L \\ p_1 &: \ L \rightarrow \{-FX + X + FX - | - FX + X + FX\} \\ p_2 &: \ X \rightarrow FX \end{aligned}$$

Production p_1 defines a leaf as a closed planar polygon. The parentheses { and } indicate that the polygon should be filled. Production p_2 linearly increases the lengths of the polygon edges. Leaves generated by this L-system were incorporated in the model of a fern shown in Figure 3.8. Note the phase effect due to the "growth" of polygons in time. A similar L-system was used to generate leaves in *Capsella bursa-pastoris* (Figure 3.10).

In practice, the tracing of polygon boundaries leads to acceptable effects only in the case of small, flat surfaces. In other cases it is more convenient to define surfaces using an underlying tree structure as a *frame*. The entire surface is then bounded by contour edges which connect endpoints of the frame segments. The difference between the three leaf shapes shown in

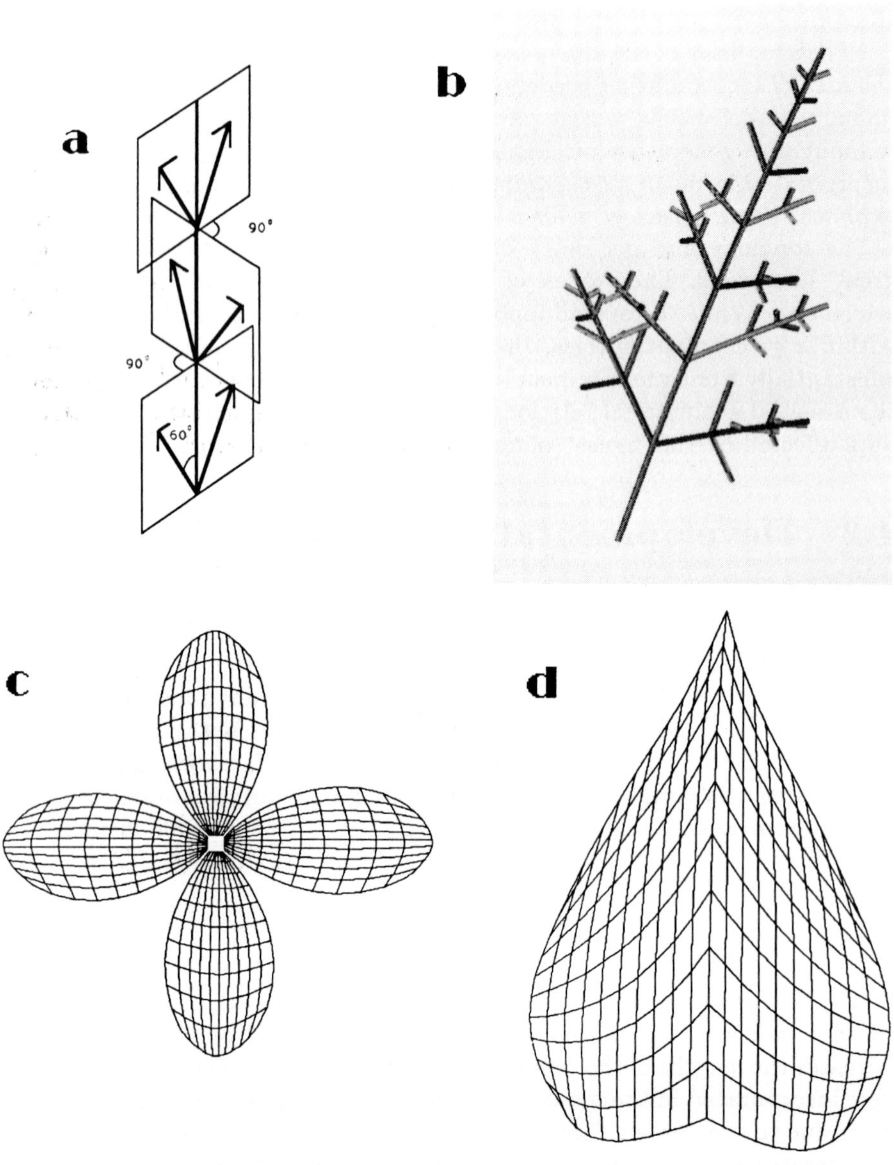

Figure 4.2: Construction of a lilac twig. (a) Geometric relationships in an inflorescence. (b) Inflorescence "skeleton" generated by an L-system. (c) Patches representing a flower. (d) Patches representing a leaf.

Figure 4.3: A lilac twig.

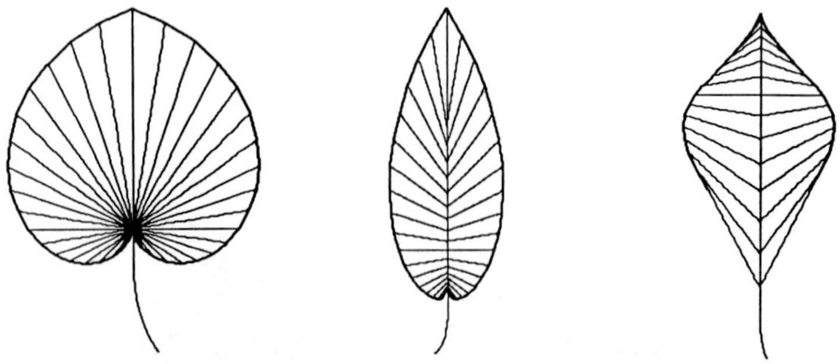

Figure 4.4: Developmental models of leaves.

Figure 4.4 results from modifying the branching angles and growth rates of axes. Specifically, the blade of the *cordate* leaf (the leftmost in Figure 4.4) was generated by the following L-system:

$$
\begin{array}{rl}
\omega: & [A][B] \\
p_1: & A \rightarrow [+A?]C\# \\
p_2: & B \rightarrow [-B?]C\# \\
p_3: & C \rightarrow IC
\end{array}
$$

The axiom contains symbols A and B which generate the left-hand side and the right-hand side of the blade. Each of the productions p_1 and p_2 creates a sequence of axes starting at the leaf base and gradually diverging from the midrib. Production p_3 increases the axis lengths. The axes close to the midrib are the longest since they were created first (thus, the leaf shape is yet another manifestation of the phase effect). Pairs of symbols ? and # indicate the endpoints of the inserted edges. The following string represents the left-hand side of the leaf after four derivation steps:

$$
[+[+[+[+A\,\underbrace{?]C\#}\,\underbrace{?]IC\#}\,\underbrace{?]IIC\#}\,\underbrace{?]IIIC\#}
$$

At this stage four edges are inserted between the vertices denoted by symbols ? and #, as indicated by the braces. The first edge has zero length, the second is collinear with an axis, and the remaining two complete triangles. An entire developmental sequence of a cordate leaf is shown in Figure 4.5.

The frame-based approach can be extended to three-dimensional organs. Figure 4.6 reveals construction of the flowers of the lily-of-the-valley from Figure 3.7. The L-system generates a supporting framework composed of five curved lines which spread radially from the flower base and are connected

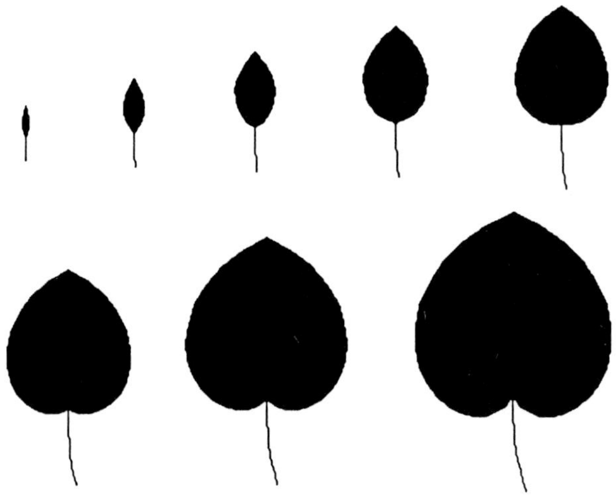

Figure 4.5: Development of a cordate leaf.

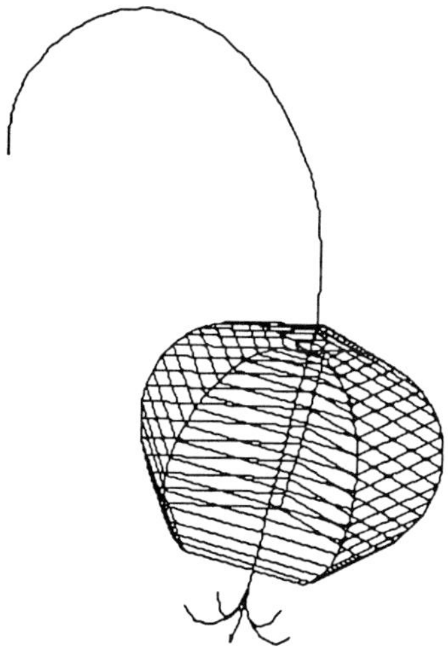

Figure 4.6: Structure of a lily-of-the-valley flower.

by a web of inserted edges. The flower is represented as a polygon mesh consisting of trapezoids bounded by two "regular" and two inserted edges.

Another developmental approach to leaf modelling was recently proposed by Lienhardt and Françon [47,48]. While they also use two types of edges to specify leaf structures, the development is not described in terms of L-systems.

Chapter 5

Models of cell layers

Although the developmental surface models discussed in the previous chapter allow for specifying closed polygons, the basic supporting structure is still limited to trees (in the graph-theoretic sense). An extension of L-systems to *maps* [104] (planar graphs with cycles which partition the plane into regions) is termed *map L-systems* [62,90,16,55,53].

Map OL-system

In the context-free case (a map OL-system), the predecessor of each production represents a single *directed* edge. The successor is a bracketed string which, in addition to letters denoting edge labels and brackets, may contain symbols $+, -$ and ˜. The symbols $+$ and $-$ specify whether a branch should be placed to the left or to the right of the main axis, but, in contrast to the turtle interpretation, do not correspond to any particular value of the branching angle. The symbol ˜ is used to indicate edge direction. The following rules apply:

- An edge which belongs to the main axis of the production successor has the same direction as the predecessor if it is not preceded by a tilde (˜); otherwise it has the opposite direction.
- A branch edge is directed away from the main axis if it is not preceded by a ˜; otherwise it is directed towards the main axis.

Derivation

A derivation step in a map L-system consists of two phases:

- In the first phase, each edge in the map is replaced by its successor according to the production set;
- In the second phase, pairs of matching branches are connected together.

The second step does not have a counterpart in the L-systems described previously and requires further explanation. Two single-edge branches match if they satisfy the following three conditions:

- they enter the same region,
- they have the same label, and
- one branch is oriented away from the main axis, while the other is oriented towards the main axis.

A pair of matching branches is connected together to form a single edge. A branch without a match is discarded from the map prior to the subsequent derivation step.

An example of a map L-system is given in Figure 5.1. It presents a production set in bracketed string form, the equivalent graph representation of these productions, the axiom (square *abcd*) and the results of four derivation steps. In the first step the two phases of edge rewriting and branch connection are distinguished.

Interpretation

In order to generate images of maps, appropriate geometric interpretation rules must be specified. The maps shown in Figure 3.1 were produced according to the rules proposed by Siero, Rozenberg and Lindenmayer [90]:

- The edges are represented by straight lines.
- The starting map is represented by a regular polygon with the desired number of edges.
- Each production subdivides an edge into segments of equal length; this subdivision determines the position of new vertices which remain in the same place in subsequent maps.
- The position of an edge resulting from a branch connection are determined by the vertices at its endpoint, as specified by the previous rule.

In the biological context, direct application of interpretation rules by Siero *et al.* tends to produce maps which, although topologically correct, contain "cells" with shapes seldom observed in nature (Figure 5.2a). A modification of interpretation rules proposed by de Does and Lindenmayer [16], lends itself to more realistic images. The essential idea is to place each interior vertex of the map in the center of gravity of its neighbours (Figure 5.2b). This adjustment of vertex positions has a sound biological justification: it minimizes hypothetical forces acting along cell edges [16], thus bringing the entire structure to a state of minimum energy. The vertices lying on the outside perimeter of the map are repositioned using a separate algorithm to prevent the entire structure from collapsing. A further step towards realism,

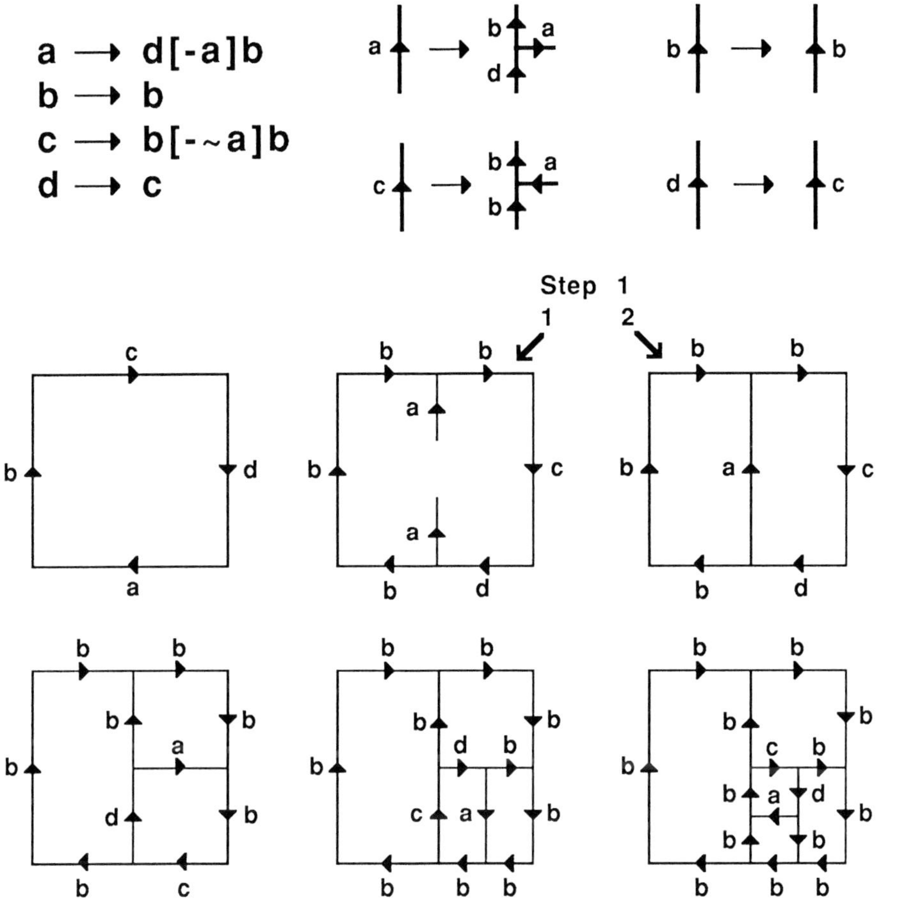

Figure 5.1: Example of a map L-system.

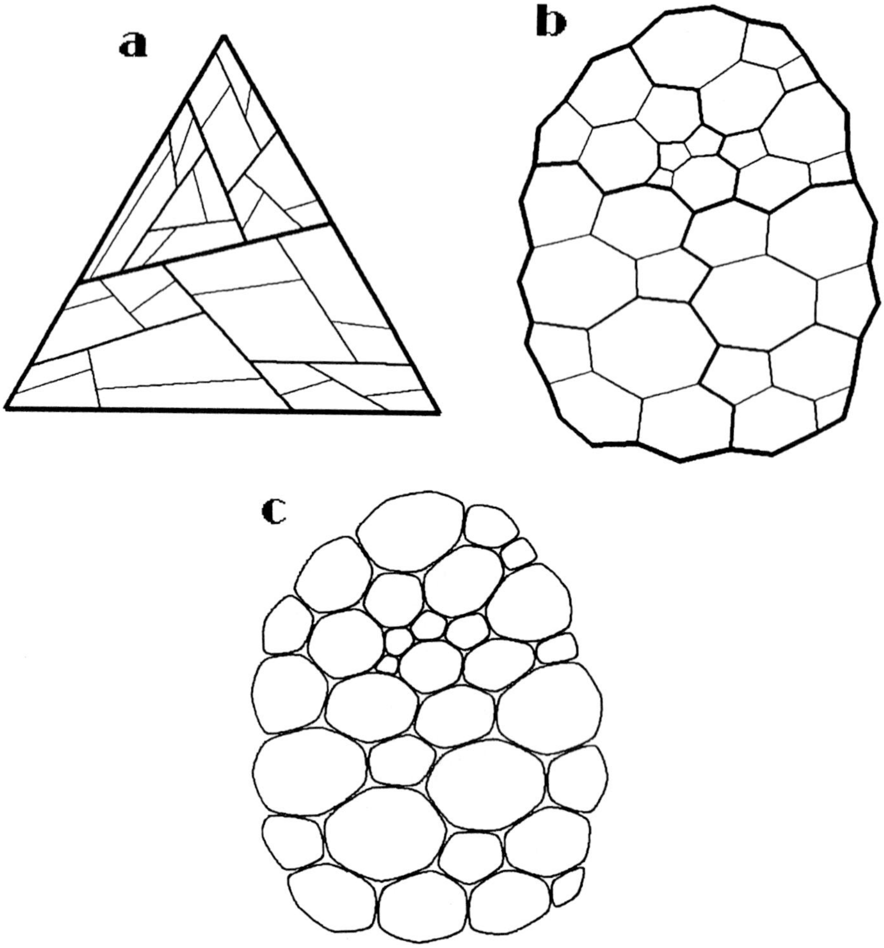

Figure 5.2: A cellular layer modelled using a map L-system. (a) Vertices not moved. (b) Vertices placed near the center of gravity of their neighbors. (c) Beta-spline approximation of the map (b). The line width in (a) and (b) is proportional to the edge age.

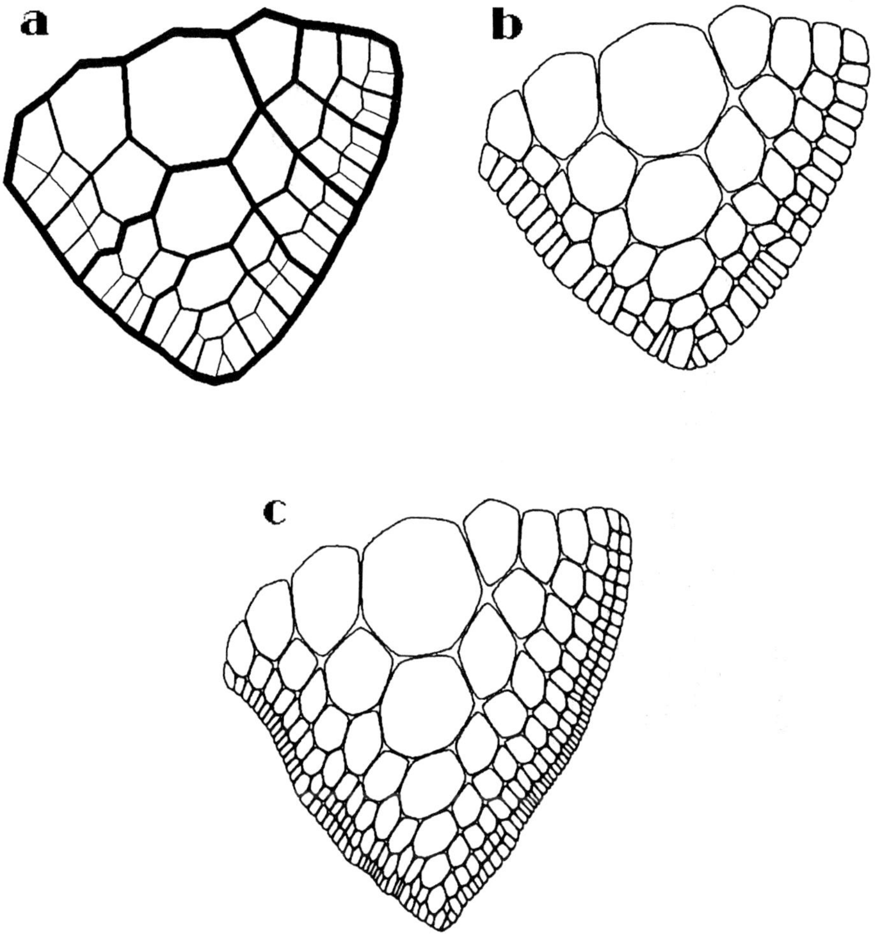

Figure 5.3: A leaf of *Phascum cuspidatum* modelled using a map L-system. (a) Vertices placed near the center of gravity of their neighbors, derivation length $n = 10$. (b) Beta-spline approximation of the structure (a) after one more derivation step. (c) The structure after two more derivation steps.

recently implemented by Dave Fracchia, consists of approximating each cell by Beta-splines [6] which use map vertices as the control points (Figure 5.2c).

Phascum

The second example of a cellular structure modelled using a map L-system is shown in Figure 5.3. The images represent young leaves of the moss *Phascum cuspidatum.*

The map L-systems used to generate Figures 5.2 and 5.3 were developed by Martin de Boer. Their listings are given below.

5.2

```
derivation length:  5
axiom:  gca
a --> c[-~a]x[+~b]g
b --> v[-b]x[+~b]v
c --> d
d --> e[-b]x[+b]e
e --> f
f --> c[-~b]x[+~b]c
g --> h[-a]i
h --> j
i --> x[+b]k
j --> l[-b]m
k --> n[-~b]o
l --> p
m --> x[+~b]p
n --> q[+~b]x
o --> q
p --> r[-~b]s
q --> t[-b]u
r --> j
s --> x[+b]j
t --> k
u --> x[+b]k
v --> ~b
x --> x
```

5.3

```
derivation length:  8
axiom:  c~p~j
a --> d[-j]c
c --> [+~y]e
d --> d
e --> f
f --> d[-y]c
j --> d[+a]~k
k --> m[+y]d
m --> n
n --> k[-~y]d
p --> q[+~a]r
q --> r[+~j]p
r --> s
s --> t
t --> r[+u]r
u --> v
v --> w[-~y][+y]d
w --> u
y --> d[+~u]d
```

Chapter 6

Other applications of L-systems

Graphical applications of L-systems are not limited to fractals and plants. In this chapter we outline some of the lesser known ones.

6.1 Patterns and tilings

Many geometric patterns and tilings can be generated using L-systems. For example, Dekking applied L-systems to produce Penrose tilings [17,18]. Two other tilings created by L-systems are shown in Figure 6.1. The problem of describing patterns and tilings using L-systems is still a largely unexplored one. From this perspective, the example-filled book by Grünbaum and Shephard [33] is a gold mine of open problems and puzzles.

6.2 Kolam patterns

6.2.1 What is a kolam?

In villages in the South of India, women decorate the courtyards in front of their homes with traditional designs called *kolam*. The art of kolam is of ancient origin; it is estimated to have developed as early as 5,000 years ago. Vivid descriptions of the beauty of kolams and the techniques for working them out can be found in Sanskrit works.

Many kolam designs are geometric patterns formed by means of interleaving straight or curved lines. They can be of different sizes and complexities. Designs as big as 3m x 3m, with total line length of hundreds of meters, can be encountered. One method of covering a large area with drawings is to repeat a small design many times. This is seldom done. A more elegant approach is to connect small elements into sophisticated structures. In

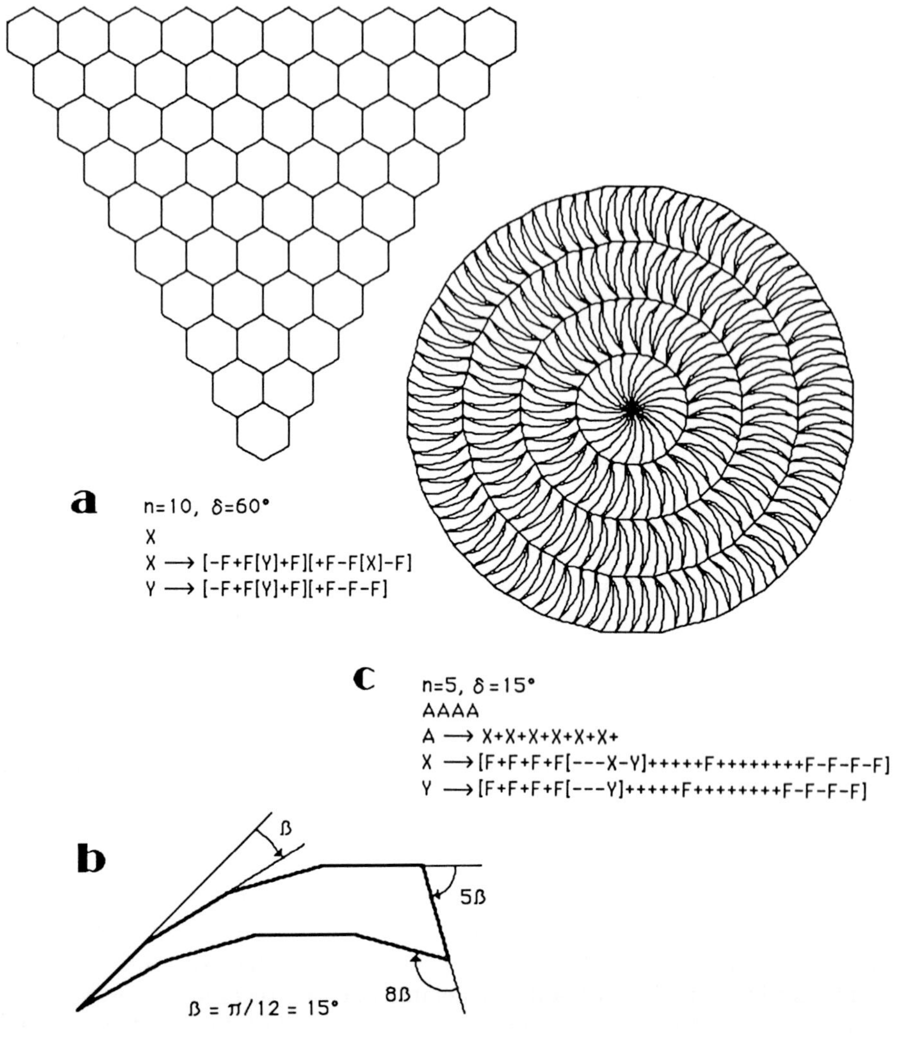

Figure 6.1: Examples of tilings generated by L-systems. (a) A hexagonal tiling. (c) A spiral tiling using the tile (b).

mathematical terms, this amounts to exploring the recursive structure of a family of increasingly more complex patterns.

6.2.2 Kolams and L-systems

The first formal model for describing kolam patterns was proposed by Siromoney, Siromoney and Krithivasan [91]. According to this model, kolam patterns can be generated by a variant of array grammars. Recently, we have noticed that L-systems with turtle interpretation offer a simpler and more intuitive method for generating kolam patterns and describing their families.

Generally, kolam patterns can be divided into three classes: non-recursive designs, recursive designs with an exponential growth, and recursive designs with a polynomial growth. The term "growth" refers to the increase in the number of primitive elements (such as vectors or spline control points) present in subsequent patterns belonging to the same family. A kolam family is understood as the sequence of patterns produced by a given L-system in derivations of length $0, 1, 2, 3, \ldots$.

Non-recursive kolams are the only members of one-element families. They can be generated by trivial L-systems in which the axiom describes the entire pattern. From the viewpoint of image generation techniques, non-recursive kolams are of little interest.

6.2.3 Kolams with exponential growth

Snake

The L-systems which generate recursive kolam patterns with an exponential growth are remarkably simple. For example, consider the variant of the *snake* kolam shown in Figure 6.2. It can be divided into two congruent parts by a diagonal line passing through opposite corners of the pattern. The recursive structure of the bottom right part is analyzed in Figure 6.3. The main component of the design is the open polygon **a**. The design starts with two line segments meeting at the right angle **b**. These lines are disconnected at the vertex marked with a dot to accommodate insertion of the polygon **a**. The polygon **c** results. The subsequent shape **d** is obtained by properly inserting polygon **a** in all marked vertices of **c**. Thus, the recursion consists of concurrently inserting the predefined shape **a** in all marked vertices of the polygon from the previous level. It is obvious that the number of polygon edges grows exponentially with the recursion depth.

The L-system generating the family of snake kolams can be obtained in a straightforward way by coding the inserted polygon **a** and the initial angular

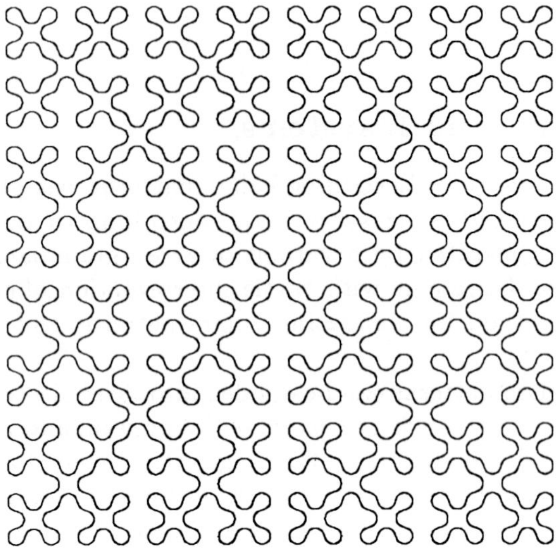

Figure 6.2: The *snake* kolam.

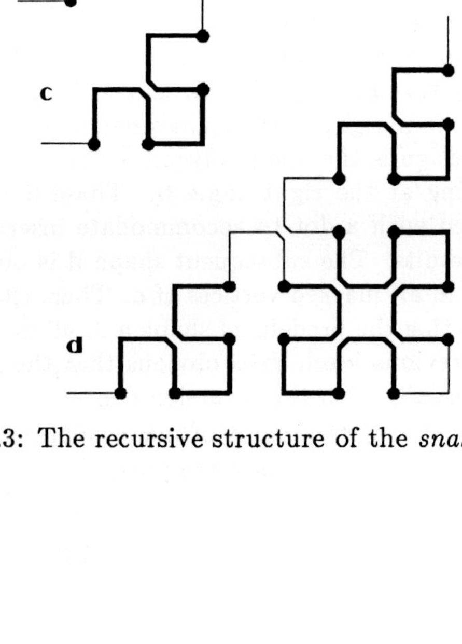

Figure 6.3: The recursive structure of the *snake* kolam.

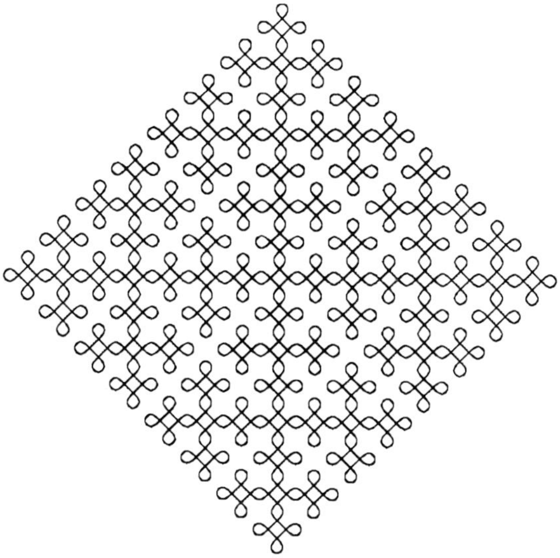

Figure 6.4: *Anklets of Krishna.*

shape **b** using turtle symbols:

$$\begin{array}{ll} \omega: & f + Xf + f + Xf \\ p: & X \rightarrow Xf - f - f + Xf + f + Xf - f - f + X \end{array}$$

The angle increment δ is equal to 90°. In order to render the curvatures of the original design, the consecutive positions of the turtle are used as the control points for a B-spline approximation (cf. Figure 2.10). The snake shown in Figure 6.2 was obtained using a derivation of length 4. Note the similarity between the snake kolam and the Sierpiński space-filling curve (Figure 2.7c). It is fascinating to discover a fractal curve in folk art!

Another kolam pattern with exponential growth, called the *anklets of Krishna*, is shown in Figure 6.4. Its structure is close to that of the snake and can be analyzed in a similar way. Figure 6.4 was generated using the following L-system: *Anklets of Krishna*

$$\begin{array}{ll} \omega: & -X - -X \\ p: & X \rightarrow XfX - -XfX \end{array}$$

The derivation length n was equal to 5 and the angle increment δ was equal to 45°.

In a modification of the *anklets of Krishna* shown in Figure 6.5a, the basic structural elements (bold squares) are separated from each other and *Bag of candies*

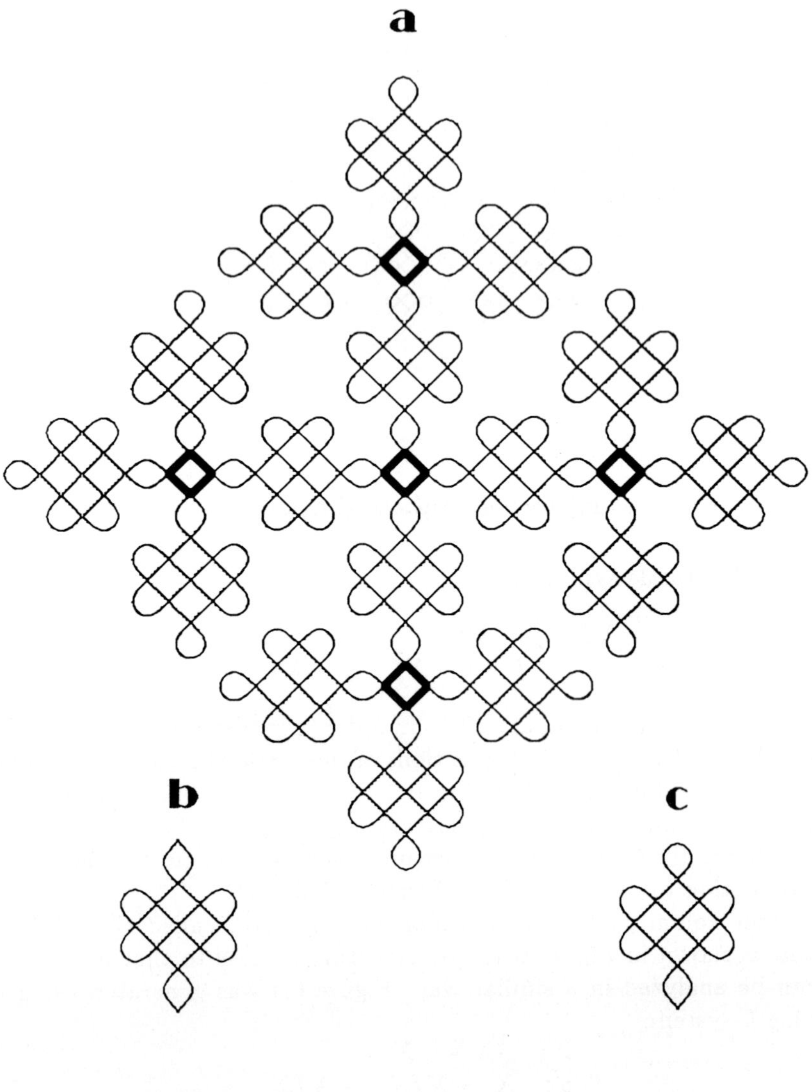

Figure 6.5: (a) The *bag of candies* kolam and (b, c) its decorative elements.

ornamented using decorative elements **b** and **c**. The pattern is described by the following table L-system:

Table 1

$$\begin{array}{ll} \omega: & -X--X \\ p_1: & X \to XfX--XfX \end{array}$$

Table 2

$$\begin{array}{ll} p_2: & X \to BfA--BfA \\ p_3: & A \to f++ffff--f--ffff++f++ffff--f \\ p_4: & B \to f--ffff++f++ffff--f--ffff++f \end{array}$$

Table 1 (production p_1) is applied in the first $n-2$ derivation steps. Table 2 is then used to insert the decorative elements. To make the structure legible, Figure 6.5a was generated with only 3 derivation steps. The angle increment δ was equal to 45°.

6.2.4 Kolams with polynomial growth

Mango leaves

The *mango leaves* kolam shown in Figure 6.6a was generated by the following L-system:

$$\begin{array}{ll} \omega: & A---A \\ p_1: & A \to f-F+Z+F-fA \\ p_2: & Z \to F-FF-F--[--Z]F-FF-F--F-FF-F-- \end{array}$$

The derivation length n was equal to 7 and the angle increment δ was equal to 60°. The axiom defines the structure as consisting of two parts, originating from the letters A and symmetric with respect to the kolam center. Production p_1 describes the formation of a sequence of wedges **b** which form the axis of the structure. Attached to each wedge is a branch **c**, generated from the symbol Z by production p_2. The number of branch elements increases in each derivation step. Since the branches near the kolam base are created first, they are longer than those situated in the higher portions of the structure. The structure exhibits polynomial growth, with the number of line segments proportional to the square of the derivation length.

Other examples of kolam patterns with polynomial growth functions are shown in Figures 6.7 and 6.8.

Kolam, fractals and plants.

It is interesting to notice the relationship between kolam patterns, fractals and plants. On one hand, kolam patterns with exponential growth are closely related to space-filling curves. On the other hand, the patterns with polynomial growth exhibit phase effects common to many plants. As in the case of plants, it is easier to design an L-systems capturing the entire sequence a

b

Figure 6.8: Further examples of kolam patterns with polynomial growth functions: (a) *Kooja* and (b) *Scissors*.

of kolam patterns forming a given family than attempt to characterize a particular "developmental stage" without referring to the growth dynamics.

More results on the formal structure of kolam patterns were recently published by G. and R. Siromoney [92].

6.3 Fractal music

In these notes we have concentrated on graphical applications of L-systems. However, at least one other application deserves attention: the automatic generation of music. The relation to graphics is, in fact, very close: an L-system is used to generate a fractal curve, which is subsequently interpreted as a sequence of notes. We will present this concept by referring to an example (Figure 6.9).

Suppose that the Hilbert curve (a) is traversed in the direction indicated by the arrow and the consecutive horizontal line segments are interpreted as notes. The pitch of each note corresponds to the y-coordinate of the segment, and the note duration is proportional to the segment length. The resulting sequence of notes forms a simple score shown in Figures 6.9b and c. Naturally, any curve consisting of horizontal and vertical segments can be interpreted in a similar way.

In the above example it is assumed that the notes belong to the C major scale and the first note is C. In general it is convenient to use a lookup table which allows for specifying an arbitrary mapping of y coordinates into note pitches. For example, interesting scores can be obtained by interpreting the Peano curve in the pentatonic scale, and by interpreting the dragon curve in the blues scale.

The musical scores obtained using L-systems are relatively complex (in spite of the simplicity of the underlying productions) but they also have a legible internal structure (they do not give the impression of notes accidentally put together). The principle of data base amplification turns out to be as attractive in the domain of musical applications as it is in graphics.

Another approach to music generation using L-systems is described in [44].

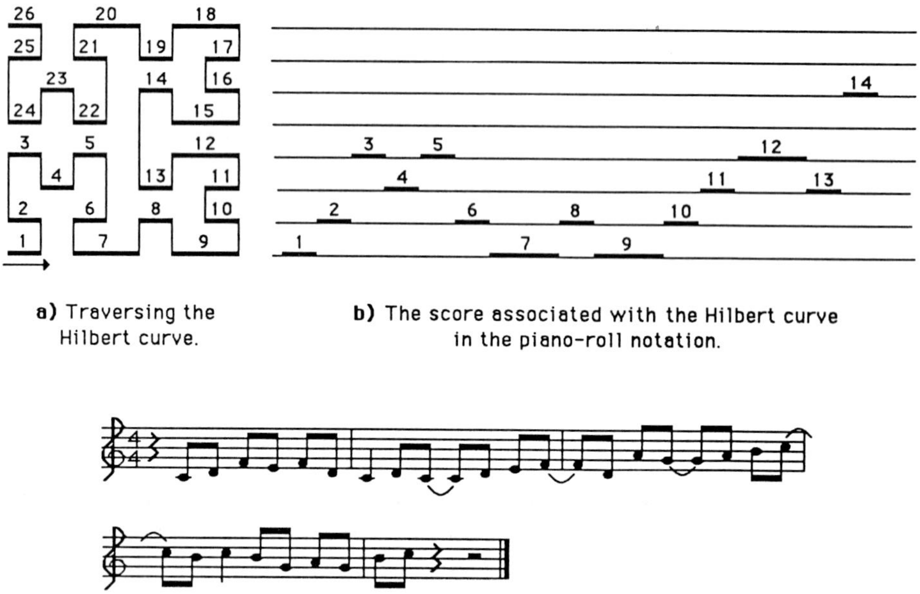

a) Traversing the Hilbert curve.

b) The score associated with the Hilbert curve in the piano-roll notation.

c) The score associated with the Hilbert curve in the common musical notation.

Figure 6.9: Musical interpretation of a curve generated by an L-system.

Chapter 7

A guide to the references

This chapter presents bibliographical remarks which focus on plant modelling and growth simulation for computer graphics purposes. An emphasis is put on the applications of L-systems.

7.1 General

Stevens [99] presents various biological structures from a mathematical perspective. D'Arcy Thompson [102] and Jean [40] emphasize the relationship between growth and form; this relationship is the cornerstone of the developmental approach to plant modelling. Lindenmayer [59] discusses the role of mathematical theories in biology.

7.2 Surveys

In his SIGGRAPH tutorial notes [95], A. R. Smith discusses the role of rewriting concepts in graphics, specifically as applied to the modelling of plants. The role of L-systems is emphasized and details of Smith's programs *Gene* and *mktree* are given. An annotated graftal bibliography is included.

The survey by Fournier [25] is a general introduction to the field of natural phenomena modelling, with an emphasis on fractals and on the modelling of biological structures.

In [51,50,52] and [55] Lindenmayer surveys the development of ideas related to L-systems in the biological context.

Macdonald [71] presents a good introduction to L-systems from a biological perspective.

7.3 Theory of L-systems

A lucid introduction to the theory of L-systems is included in Salomaa's textbook on formal languages [89]. Herman and Rozenberg wrote the first (and still very useful) book entirely devoted to L-systems [37]. A more recent monograph by Rozenberg [88] puts emphasis on OL-systems. Many theoretical results are contained in Vitányi's Ph.D. thesis [106]. Collections of papers [61] and [87] combine formal aspects of L-systems with biological applications.

Two extensive bibliographies of L-systems were published in 1977 [86] and 1981 [85].

As the number of theoretical papers on L-systems is estimated at approximately one thousand, we mention here only those directly applicable to plant modelling. The original paper on L-systems by Lindenmayer [54] introduces the notions of context-free, context-sensitive and bracketed L-systems. Eichhorst and Savitch [21] and Yokomori [112] present a definition of stochastic L-systems which is useful in the modelling of specimen-to-specimen variation [78,81]. Jürgensen and Lindenmayer [42] analyze an inference problem which consists of finding a DOL-system generating an observed sequence of tree structures. Wood and Culik [111,110,14] introduce the concept of time-delayed L-systems which may be applicable to the simulation of continuous growth. Vitányi [105] attempts to obtain the sigmoidal growth curve (describing the growth pattern of most plants) using L-systems; however, the concept of slowing down the progress of time at a certain age seems artificial. Jones and Skyum [41] review results on computation and complexity of L-systems; these results can be applied to estimate the complexity of control mechanisms in plants.

7.4 Geometrical interpretation of L-systems

L-systems were conceived for the purpose of describing the development of plants, but this description was originally confined to the topological level. In order to present plant development using computer-generated images, it is necessary to specify a geometrical interpretation of L-systems. The first approaches — by Hogeweg and Hesper [38], and Frijters and Lindenmayer [30] — used a constant branching angle throughout the entire structure. Szilard and Quinton [101] introduced the idea of using symbols generated by an L-system to specify directions and angles between line segments. Further mathematical results related to this approach were obtained by Dekking [17,18]. Siromoney and Subramanian [93] applied chain coding [26] to generate space-filling curves using L-systems. Maurer, Rozenberg and Welzl [73] and Sudborough and Welzl [100] analyzed formal properties of pictures generated by Chomsky grammars under the chain interpretation; it would be most interesting to extend these results to L-systems. Prusinkiewicz [82,81] explored turtle geometry (described in [1]) as the basis for interpreting strings generated by L-systems. Smith [96] initiated a unified mathematical theory of rewriting systems for graphics purposes.

7.5 Biological applications of L-systems

The first program for simulating plant development with L-systems, called CELIA, was designed by Baker, Herman and Liu [5,36,49]. The first computer-generated plots of plant structures are included in the papers by Hogeweg and Hesper [38], and Frijters and Lindenmayer [30]. In a pair of papers [27,28], Frijters discusses delays, sequences of stages and developmental switches as a basis for the simulation of inflorescence development. Developmental models incorporating interactive control mechanisms were introduced by Lindenmayer [58] and refined by Janssen and Lindenmayer [39]. *Mycelis muralis* was used as an example.

In [29], Frijters and Lindenmayer introduce the notion of paracladial relations to describe self-similarities in branching structures on the topological level. An extension of this notion to leaf structures is presented by Lindenmayer [57]. A simple L-system describing the topology of *Delphinium ajacis* leaves is given, but the addition of geometric parameters is not a trivial problem in this case. In contrast, stochastic L-systems applied by Nishida to model development of shoots of Japanese cypress [78] lend themselves easily to geometric interpretation. A notation related to L-systems is introduced for the purpose of inflorescence description by Robinson [84].

7.6 Synthesis of realistic plant images

7.6.1 Methods based on L-systems

The first realistic images of computer-generated plants were produced by Smith [94] using 2L-systems obtained by Hogeweg and Hesper [38]. In [98], Smith presents his method in more detail and illustrates it using many realistic plant images. There is also a video tape showing these plants in development [97].

Prusinkiewicz applied L-systems with turtle interpretation to model two-dimensional [82] and three-dimensional [81] plants. Stochastic L-systems are used to achieve specimen-to-specimen variation within a species, and bicubic patches are integrated with the L-system-based model. L-systems generating several plant images are specified in detail.

Beyer and Friedell [7] claim to have invented a "new theory of scene modelling" while ignoring most of the previous work in the field, including the 1984 paper by Smith [98]. However, the extension of L-systems to graphs with cycles seems to be different from the map L-systems of Lindenmayer and the double-wall L-systems of the Lücks (Section 7.7).

7.6.2 Other synthesis methods

An early approach to the algorithmic generation of branching structures for computer graphics purposes was developed by Kawaguchi [43]. Aono and Kunii [3,2] define tree topology by recursive algorithms; variation between species is achieved primarily by manipulating branching angles. Bloomenthal [8] uses splines to model branches, carefully models the branching area, and uses texture-mapping to render leaves and bark. In a subsequent paper [9], he explores the branching area in more detail. A related approach to the modelling of trees is presented by Oppenheimer [79]. Eyrolles [22] generates realistic two-dimensional tree silhouettes generated using a stochastic technique based on Horton-Strahler analysis (a classic method for analyzing river systems). Reeves and Blau [83] apply particle systems to produce impressionistic images of trees, forests and flower fields. Lienhardt and Françon [47,48] discuss developmental models of surfaces, with an emphasis on leaves.

7.7 Map L-systems

Graph-rewriting systems can be divided into two categories, those which replace nodes and those which replace edges. So far, only the edge-rewriting systems have found biological applications. The theoretical background of edge rewriting is presented by Habel and Kreowski [34].

Two approaches to the rewriting of parallel graphs with cycles (used to model cellular structures) are known as map L-systems and double-wall L-systems. The concepts preceding that of map L-systems are described in papers by Culik, Lindenmayer and Wood [12,60,13]. Lindenmayer and Rozenberg are the authors of the original paper on map L-systems [62]. Detailed examples of the operation of map L-systems are given by Siero, Rozenberg and Lindenmayer [90], and de Does and Lindenmayer [16]. They also introduce several graphical interpretations of maps. An example of a three-dimensional L-system, or cellwork, is presented by Lindenmayer [56]. His paper also solves the geometric problem of calculating edge lengths after cell subdivision. A recent survey of map L-systems is presented by [53]. A different approach to the definition of cellular structures was developed by J. Lück and H. B. Lück, and termed double-wall L-systems [68,69,67,65,70,63,66]. Their most recent paper is [64].

Bibliography

[1] H. Abelson and A. A. diSessa. *Turtle geometry.* M.I.T. Press, Cambridge, 1982.

[2] M. Aono and T. L. Kunii. Botanical tree image generation, Video tape, IBM, Japan, Tokyo, 1985.

[3] M. Aono and T. L. Kunii. Botanical tree image generation. *IEEE Computer Graphics and Applications*, 4(5):10–34, 1984.

[4] J. W. Backus. The syntax and semantics of the proposed international algebraic language of the Zurich ACM-GAMM conference. In *Proc. Intl. Conf. on Information Processing*, pages 125–132, UNESCO, 1959.

[5] R. Baker and G. T. Herman. Simulation of organisms using a developmental model, parts I and II. *Int. J. of Bio-Medical Computing*, 3:201–215 and 251–267, 1972.

[6] B. A. Barsky. *The Beta-spline: a local representation based on shape parameters and fundamental geometric measures.* PhD thesis, Department of Computer Science, University of Utah, 1981.

[7] T. Beyer and M. Friedell. Generative scene modelling. In *Proceedings of EUROGRAPHICS '87*, pages 151–158, 571, 1987.

[8] J. Bloomenthal. Modeling the mighty maple. *Computer Graphics*, 19(3):305–311, 1985.

[9] J. Bloomenthal. *Polygonization of implicit surfaces.* Report CSL-87-2, Xerox Corporation, Palo Alto, CA, 1987.

[10] N. Chomsky. Three models for the description of language. *IRE Trans. on Information Theory*, 2(3):113–124, 1956.

[11] V. Claus, H. Ehrig, and G. Rozenberg, editors. *Graph grammars and their application to computer science; First International Workshop. Lecture Notes in Computer Science 73*, Springer-Verlag, Berlin, 1979.

[12] K. Culik II and A. Lindenmayer. Parallel graph generating and graph recurrence systems for multicellular development. *Int. J. General Systems*, 3:53–66, 1976.

[13] K. Culik II and D. Wood. A mathematical investigation of propagating graph OL-systems. *Information and Control*, 43:50–82, 1979.

[14] K. Culik II and D. Wood. Speed-varying OL systems. *Information Sciences*, 14:161–170, 1978.

[15] C. Davis and D. E. Knuth. Number representations and dragon curves. *J. of Recreational Mathematics*, 3:66–81, 133–149, 1970.

[16] M. de Does and A. Lindenmayer. Algorithms for the generation and drawing of maps representing cell clones. In H. Ehrig, M. Nagl, and G. Rozenberg, editors, *Graph grammars and their application to computer science; Second International Workshop*, pages 39–57, Springer-Verlag, Berlin, 1983. Lecture Notes in Computer Science 153.

[17] F. M. Dekking. Recurrent sets. *Advances in Mathematics*, 44(1):78–104, 1982.

[18] F. M. Dekking. *Recurrent sets: A fractal formalism.* Report 82-32, Delft University of Technology, 1982.

[19] H. Ehrig, M. Nagl, A. Rosenfeld, and G. Rozenberg, editors. *Graph grammars and their application to computer science; Third International Workshop. Lecture Notes in Computer Science 291*, Springer-Verlag, Berlin, 1987.

[20] H. Ehrig, M. Nagl, and G. Rozenberg, editors. *Graph grammars and their application to computer science; Second International Workshop. Lecture Notes in Computer Science 153*, Springer-Verlag, Berlin, 1983.

[21] P. Eichhorst and W. J. Savitch. Growth functions of stochastic Lindenmayer systems. *Information and Control*, 45:217–228, 1980.

[22] G. Eyrolles. *Synthèse d'images figuratives d'arbres par des méthodes combinatoires.* PhD thesis, Université de Bordeaux I, 1986.

[23] J. Feder. Languages of encoded line patterns. *Information and Control*, 13:230–244., 1968.

[24] James D. Foley and Andries Van Dam. *Fundamentals of Interactive Computer Graphics. The Systems Programming Series*, Addison-Wesley Publishing Company, Reading, Massachusetts, 1982.

[25] A. Fournier. Prolegomenon. SIGGRAPH '87 Course Notes on the Modeling of Natural Phenomena, 1987.

[26] H. Freeman. On encoding arbitrary geometric configurations. *IRE Trans. Electron. Comput*, 10:260–268, 1961.

[27] D. Frijters. Mechanisms of developmental integration of *Aster novae-angliae* L. and *Hieracium murorum* L. *Annals of Botany*, 42:561–575, 1978.

[28] D. Frijters. Principles of simulation of inflorescence development. *Annals of Botany*, 42:549–560, 1978.

[29] D. Frijters and A. Lindenmayer. Developmental descriptions of branching patterns with paracladial relationships. In A. Lindenmayer and G. Rozenberg, editors, *Automata, languages, development*, pages 57–73, North-Holland, Amsterdam, 1976.

[30] D. Frijters and A. Lindenmayer. A model for the growth and flowering of *Aster novae-angliae* on the basis of table (1, 0) L-systems. In G. Rozenberg and A. Salomaa, editors, *L Systems*, pages 24–52, Springer-Verlag, Berlin, 1974. Lecture Notes in Computer Science 15.

[31] K. S. Fu. Syntactic (linguistic) pattern recognition. In K. S. Fu, editor, *Digital pattern recognition*, pages 95–134, Springer-Verlag, Berlin - Heidelberg - New York, 1980.

[32] S. Ginsburg and H. G. Rice. Two families of languages related to ALGOL. *J. ACM*, 9(3):350–371, 1962.

[33] Branko Grünbaum and G. C. Shephard. *Tilings and Patterns.* W. H. Freeman and Company, New York, 1987.

[34] A. Habel and H.-J. Kreowski. On context-free graph languages generated by edge replacement. In H. Ehrig, M. Nagl, and G. Rozenberg, editors, *Graph grammars and their application to computer sci-*

ence; Second International Workshop, pages 143–158, Springer-Verlag, Berlin, 1983. Lecture Notes in Computer Science 153.

[35] F. Hallé, R. A. A. Oldeman, and P. B. Tomlinson. *Tropical trees and forests: an architectural analysis*. Springer-Verlag, Berlin, 1978.

[36] G. T. Herman and W. H. Liu. The daughter of CELIA, the French flag, and the firing squad. *Simulation*, 21:33–41, 1973.

[37] G. T. Herman and G. Rozenberg. *Developmental systems and languages*. North-Holland, Amsterdam, 1975.

[38] P. Hogeweg and B. Hesper. A model study on biomorphological description. *Pattern Recognition*, 6:165–179, 1974.

[39] J. M. Janssen and A. Lindenmayer. Models for the control of branch positions and flowering sequences of capitula in *Mycelis muralis* (L.) dumont (Compositae). *New Phytologist*, 105:191–220, 1987.

[40] R. V. Jean. *Mathematical approach to pattern and form in plant growth*. John Wiley and Sons, New York, 1984.

[41] N. D. Jones and S. Skyum. Complexity of some problems concerning L-systems. *Mathematical Systems Theory*, 13:29–43, 1979.

[42] H. Jürgensen and A. Lindenmayer. Inference algorithms for developmental systems with cell lineages. *Bulletin of Mathematical Biology*, 49(1):93–123, 1987.

[43] Y. Kawaguchi. A morphological study of the form of nature. *Computer Graphics*, 16(3):223–232, 1982.

[44] P. S. Langston. (201) 644-2332 or Eedie & Eddie on the wire — An experiment in music generation. Bell Laboratories, New Jersey, 1986.

[45] R. S. Ledley. High-speed automatic analysis of biomedical pictures. *Science*, 146(3641):216–223, 1964.

[46] R. S. Ledley et al. FIDAC: Film input to digital automatic computer and associated syntax-directed pattern-recognition programming system. In J. T. Tippet et al., editors, *Optical and electro-optical information processing*, pages 591–613, M.I.T. Press, Cambridge, 1965.

[47] P. Lienhardt. *Modélisation et évolution de surfaces libres*. PhD thesis, Université Louis Pasteur, Strasbourg, 1987.

[48] P. Lienhardt and J. Françon. *Synthèse d'images de feuilles végétales.* Technical Report R-87-1, Département d'informatique, Université Louis Pasteur, Strasbourg, 1987.

[49] A. Lindenmayer. Adding continuous components to L-systems. In G. Rozenberg and A. Salomaa, editors, *L Systems*, pages 53–68, Springer-Verlag, Berlin, 1974. Lecture Notes in Computer Science 15.

[50] A. Lindenmayer. Algorithms for plant morphogenesis. In R. Sattler, editor, *Theoretical plant morphology*, pages 37–81, Leiden University Press, The Hague, 1978.

[51] A. Lindenmayer. Developmental algorithms for multicellular organisms: a survey of L-systems. *Journal of Theoretical Biology*, 54:3–22, 1975.

[52] A. Lindenmayer. Developmental algorithms: Lineage versus interactive control mechanisms. In S. Subtelny and P. B. Green, editors, *Developmental order: Its origin and regulation*, pages 219–245, Alan R. Liss, Inc., New York, 1982.

[53] A. Lindenmayer. An introduction to parallel map generating systems. In H. Ehrig, M. Nagl, A. Rosenfeld, and G. Rozenberg, editors, *Graph grammars and their application to computer science; Third International Workshop*, pages 27–40, Springer-Verlag, Berlin, 1987. Lecture Notes in Computer Science 291.

[54] A. Lindenmayer. Mathematical models for cellular interaction in development, Parts I and II. *Journal of Theoretical Biology*, 18:280–315, 1968.

[55] A. Lindenmayer. Models for multicellular development: characterization, inference and complexity of L-systems. In A. Kelmenová and J. Kelmen, editors, *Trends, techniques and problems in theoretical computer science*, pages 138–168, Springer-Verlag, Berlin, 1987. Lecture Notes in Computer Science 281.

[56] A. Lindenmayer. Models for plant tissue development with cell division orientation regulated by preprophase bands of microtubules. *Differentiation*, 26:1–10, 1984.

[57] A. Lindenmayer. Paracladial relationships in leaves. *Ber. Deutsch Bot. Ges. Bd.*, 90:287–301, 1977.

[58] A. Lindenmayer. Positional and temporal control mechanisms in inflorescence development. In P. W. Barlow and D. J. Carr, editors, *Positional controls in plant development*, University Press, Cambridge, 1984.

[59] A. Lindenmayer. Theories and observations of developmental biology. In R. E. Butts and J. Hintikka, editors, *Foundational problems in special sciences*, pages 103–118, D. Reidel Publ. Co, Dordrecht-Holland, 1977.

[60] A. Lindenmayer and K. Culik II. Growing cellular systems: Generation of graphs by parallel rewriting. *Int. J. General Systems*, 5:45–55, 1979.

[61] A. Lindenmayer and G. Rozenberg, editors. *Automata, languages, development.* North-Holland, Amsterdam, 1976.

[62] A. Lindenmayer and G. Rozenberg. Parallel generation of maps: Developmental systems for cell layers. In V. Claus, H. Ehrig, and G. Rozenberg, editors, *Graph grammars and their application to computer science; First International Workshop*, pages 301–316, Springer-Verlag, Berlin, 1979. Lecture Notes in Computer Science 73.

[63] H. B. Lück and J. Lück. Unconventional leaves (an application of map OL-systems to biology). In G. Rozenberg and A. Salomaa, editors, *The book of L*, pages 275–289, Springer-Verlag, Berlin - Heidelberg - New York - Tokyo, 1986.

[64] J. Lück, A. Lindenmayer, and H. B. Lück. Models for cell tetrads and clones in meristematic cell layers. 1988. Botanical Gazette, in press.

[65] J. Lück and H. B. Lück. Comparative plant morphogenesis founded on map and stereomap generating systems. In J. Demongeot, E. Goles, and M. Tchuente, editors, *Dynamical systems and cellular automata*, pages 111–121, Academic Press, London, 1985.

[66] J. Lück and H. B. Lück. From OL and IL map systems to indeterminate and determinate growth in plant morphogenesis. In H. Ehrig, M. Nagl, A. Rosenfeld, and G. Rozenberg, editors, *Graph grammars and their application to computer science; Third International Workshop*, pages 393–410, Springer-Verlag, Berlin, 1987. Lecture Notes in Computer Science 291.

[67] J. Lück and H. B. Lück. Generation of 3-dimensional plant bodies by double wall map and stereomap systems. In H. Ehrig, M. Nagl,

and G. Rozenberg, editors, *Graph-Grammars and Their Application to Computer Science; Second International Workshop*, pages 219–231, Springer-Verlag, Berlin, 1983. Lecture Notes in Computer Science 153.

[68] J. Lück and H. B. Lück. Proposition d'une typologie de l'organisation cellulaire des tissus végétaux. In Hervé Le Guardier and Thiebut Moulin, editors, *Actes du premier séminaire de l'Ecole de Biologie Théorique du CNRS*, pages 335–371, Ecole Nationale Superieure de Techniques Avancées, Paris, 1981.

[69] J. Lück and H. B. Lück. Sur la structure de l'organisation tissulaire et son incidence sur la morphogenèse. In Hervé Le Guardier, editor, *Actes du deuxième séminaire de l'Ecole de Biologie Théorique du CNRS*, pages 385–397, Publications de l'Université de Rouen, Abbaye de Solignac, 1982.

[70] J. Lück and H. B. Lück. Un mécanisme générateur d'hélices phyllotaxiques. In G. Benchetrit and J. Demongeot, editors, *Actes du IVe séminaire de l'Ecole de Biologie Théorique*, pages 317–330, Editions du CNRS, Paris, 1985.

[71] N. Macdonald. *Trees and networks in biological models.* J. Wiley, New York, 1983.

[72] B. B. Mandelbrot. *The fractal geometry of nature.* W. H. Freeman, San Francisco, 1982.

[73] H. A. Maurer, G. Rozenberg, and E. Welzl. Using string languages to describe picture languages. *Information and Control*, 54:155–185, 1982.

[74] G. J. Mitchison and Michael Wilcox. Rules governing cell division in *Anabaena. Nature*, 239:110–111, 1972.

[75] D. Müller-Doblies and U. Müller-Doblies. Cautious improvement of a descriptive terminology of inflorescences. Monocot newsletter 4, Institut für Biologie, Technical University of Berlin (West), 1987. 13 pp.

[76] R. Narasimhan. *A linguistic approach to pattern recognition.* Report 21, Digital Computer Laboratory, University of Illinois, Urbana, 1962.

[77] P. Naur et al. Report on the algorithmic language ALGOL 60. *Comm. ACM*, 3(5):299–314, 1960. revised in Comm. ACM 6 Nr, 1, pp. 1-17.

[78] T. Nishida. KOL-systems simulating almost but not exactly the same development — the case of Japanese cypress. *Memoirs Fac. Sci., Kyoto University, Ser. Bio*, 8:97–122, 1980.

[79] P. Oppenheimer. Real time design and animation of fractal plants and trees. *Computer Graphics*, 20(4):55–64, 1986.

[80] F. P. Preparata and R. T. Yeh. *Introduction to Discrete Structures*. Addison-Wesley Publishing Company, Reading, Massachusetts, 1973.

[81] P. Prusinkiewicz. Applications of L-systems to computer imagery. In H. Ehrig, M. Nagl, A. Rosenfeld, and G. Rozenberg, editors, *Graph grammars and their application to computer science; Third International Workshop*, pages 534–548, Springer-Verlag, Berlin, 1987. Lecture Notes in Computer Science 291.

[82] P. Prusinkiewicz. Graphical applications of L-systems. In *Proceedings of Graphics Interface '86 — Vision Interface '86*, pages 247–253, 1986.

[83] W. T. Reeves and R. Blau. Approximate and probabilistic algorithms for shading and rendering structured particle systems. *Computer Graphics*, 19(3):313–322, 1985.

[84] D. F. Robinson. A notation for the growth of inflorescences. *New Phytologist*, 103:587–596, 1986.

[85] G. Rozenberg, J. Mäenpää, and A. Salomaa. Supplementary bibliography of L systems. *Bull. Europ. Assoc. Theor. Comp. Sci. (Leiden)*, 13:64–79, 1981.

[86] G. Rozenberg, M. Penttonen, and A. Salomaa. Bibliography of L systems. *Theoretical Computer Science*, 5:339–354, 1977.

[87] G. Rozenberg and A. Salomaa, editors. *The Book of L*. Springer-Verlag, Berlin, 1986.

[88] G. Rozenberg and A. Salomaa. *The mathematical theory of L-systems*. Academic Press, New York, 1980.

[89] A. Salomaa. *Formal languages*. Academic Press, New York, 1973.

[90] P. L. J. Siero, G. Rozenberg, and A. Lindenmayer. Cell division patterns: syntactical description and implementation. *Computer Graphics and Image Processing*, 18:329–346, 1982.

[91] G. Siromoney, R. Siromoney, and K. Krithivasan. Array grammars and kolam. *Computer Graphics and Image Processing*, 63–82, 1974.

[92] G. Siromoney and R. Siromoney. Rosenfeld's cycle grammars and kolam. In H. Ehrig, M. Nagl, A. Rosenfeld, and G. Rozenberg, editors, *Graph grammars and their application to computer science; Third International Workshop*, pages 564–579, Springer-Verlag, Berlin, 1987. Lecture Notes in Computer Science 291.

[93] R. Siromoney and K. G. Subramanian. Space-filling curves and infinite graphs. In H. Ehrig, M. Nagl, and G. Rozenberg, editors, *Graph grammars and their application to computer science; Second International Workshop*, pages 380–391, Springer-Verlag, Berlin, 1983. Lecture Notes in Computer Science 153.

[94] A. R. Smith. About the cover: Reconfigurable machines. *Computer*, 11(7):3–4, 1978.

[95] A. R. Smith. Formal geometric languages for natural phenomena. SIGGRAPH '87 Course Notes on the Modeling of Natural Phenomena, 1987.

[96] A. R. Smith. *Graftal formalism notes.* Technical Memo 114, Lucasfilm Computer Division, San Rafael, CA, 1984.

[97] A. R. Smith. Grammars for generating the complexity of reality. Video tape, Lucasfilm/PIXAR, 1985.

[98] A. R. Smith. Plants, fractals, and formal languages. *Computer Graphics*, 18(3):1–10, 1984.

[99] P. S. Stevens. *Patterns in nature.* Little, Brown and Co., Boston, 1974.

[100] I. H. Sudborough and E. Welzl. Complexity and decidability for chain code picture languages. *Theoretical Computer Science*, 36:173–202, 1985.

[101] A. L. Szilard and R. E. Quinton. An interpretation for DOL systems by computer graphics. *The Science Terrapin*, 4:8–13, 1979.

[102] d'Arcy Thompson. *On growth and form.* At the University Press, Cambridge, 1952.

[103] W. Troll. *Die Infloreszenzen.* Volume I., Gustav Fischer Verlag, Stuttgart, 1964.

[104] W. T. Tutte. *Graph Theory.* Addison-Wesley Publishing Company, Reading, Massachusetts, 1982.

[105] P. M. B. Vitányi. Development, growth and time. In G. Rozenberg and A. Salomaa, editors, *The Book of L*, pages 431–444, Springer-Verlag, Berlin, 1986.

[106] P. M. B. Vitányi. *Lindenmayer systems: Structure, languages and growth functions.* Mathematical Centre, Amsterdam, 1980.

[107] H. von Koch. Une méthode géométrique élémentaire pour l'étude de certaines questions de la théorie des courbes planes. *Acta mathematica*, 30:145–174, 1905.

[108] S. Wolfram. Computer software in science and mathematics. *Scientific American*, 251(3):188–203, 1984.

[109] S. Wolfram. Some recent results and questions about cellular automata. In J. Demongeot, E. Goles, and M. Tchuente, editors, *Dynamical Systems and Cellular Automata*, pages 153–167, Academic Press, London, 1985.

[110] D. Wood. Generalized time-delayed OL languages. *Information Sciences*, 12:151–155, 1977.

[111] D. Wood. Time-delayed OL languages and sequences. *Information Sciences*, 8:271–281, 1975.

[112] T. Yokomori. Stochastic characterizations of EOL languages. *Information and Control*, 45:26–33, 1980.

[113] M. H. Zimmerman and C. L. Brown. *Trees — structure and function.* Springer-Verlag, Berlin, 1971.

List of Figures

Appendix A

Program listing

Listed below is the Macintosh version of the **pfg** program (plant and fractal generator). It can be used to reproduce most of the two-dimensional figures included in these notes. The program consists of two modules: *generate.c* and *interpret.c*. The first module generates a string according to an L-system specified in an input file. Context-sensitive productions and bracketed strings are supported. The second module interprets the resulting string as a two-dimensional black-and-white figure, according to the turtle interpretation rules. Extensions supporting modification of line width (useful for seeing signals), tropisms, filled polygons, and spline approximation of polygons are not included in this version but can be added easily.

The complete implementation of **pfg** is designed for IRIS 3130 workstations and operates in three dimensions. Models may incorporate an arbitrary number of surfaces defined using bicubic patches. The IRIS version also makes it possible to animate plant development.

A.1 generate.h

```
#define MAXPROD      50      /* maximum number of productions        */
#define MAXAXIOM     100     /* maximum length of the axiom          */
#define MAXSTR       30000   /* maximum length of the derived string */
#define MAXCHARS     256     /* the number of ASCII characters       */
#define MAXIGNORE    50      /* maximum number of ignored symbols    */

/*****************************************************************/
/* Structure "Parameter" collects various input parameters.      */
/*****************************************************************/

struct Parameter {
    char *filename;          /* input file name                      */
    int n;                   /* derivation length                    */
    int angle;               /* the angle factor                     */
    int scale;               /* the scaling factor                   */
};

typedef struct Parameter Parameter;

/*****************************************************************/
/* "Production" is a structure which contains pointers to the first */
/* characters and the lengths of: the left context, the strict   */
/* predecessor, the right context and the production successor.  */
/* The actual strings are stored linearly in the array inpStr,   */
/* separated by the '\0' character.                              */
/*****************************************************************/

struct Production {
    char *lCon;              /* the left context                     */
    int lConLen;             /* the length of the left context       */
    char *pred;              /* the strict predecessor               */
    int predLen;             /* the length of the strict predecessor */
    char *rCon;              /* the right context                    */
    int rConLen;             /* the length of the right context      */
    char *succ;              /* the successor                        */
    int succLen;             /* the length of the successor          */
};

typedef struct Production Production;
```

A.2 generate.c

```
/********************************************************************/
/*                  Plant and Fractal Generator (pfg)               */
/*                  Author: Przemyslaw Prusinkiewicz                */
/*          Copyright (C) 1988: Przemyslaw Prusinkiewicz            */
/*                                                                  */
/* This program is made available free of charge for educational   */
/* purposes only. The program cannot be sold, included entirely     */
/* or in part in commercial programs, or used for any commercial    */
/* purposes without written permission from the author.             */
/********************************************************************/

/********************************************************************/
/* Pfg was developed and tested on a Macintosh SE using Aztec C     */
/* compiler Version 3.4b. The compiling procedure is:               */
/*      cc generate.c                                               */
/*      cc interpret.c                                              */
/*      ln -o pfg generate.c interpret.c -lm -lc                    */
/********************************************************************/

/********************************************************************/
/* Pfg produces images of two-dimensional structures specified by   */
/* L-systems  with turtle interpretation. The program supports      */
/* bracketed L-systems with no context, one-sided context and       */
/* two-sided context. An example data file is given below:          */
/*                                                                  */
/*              Derivation length: 30                               */
/*              angle factor: 16                                    */
/*              scale factor: 100                                   */
/*              axiom: F1F1F1                                       */
/*              ignore: +-F                                         */
/*              0 < 0 > 0 --> 0                                     */
/*              0 < 0 > 1 --> 1[-F1F1]                              */
/*              0 < 1 > 0 --> 1                                     */
/*              0 < 1 > 1 --> 1                                     */
/*              1 < 0 > 0 --> 0                                     */
/*              1 < 0 > 1 --> 1F1                                   */
/*              1 < 1 > 0 --> 1                                     */
/*              1 < 1 > 1 --> F0                                    */
/*              * < + > * --> -                                     */
/*              * < - > * --> +                                     */
/*                                                                  */
/* The "angle factor" determines the angle increment:               */
/* delta = 360 degrees / (angle factor).                            */
/*                                                                  */
/* The "scale factor" determines the size of the resulting          */
/* image. In principle, 100 denotes a full-screen image and         */
/* 0 denotes an image reduced to a single pixel.                    */
```

```
/* However, in order to reduce distortion, the image           */
/* is always scaled in such a way that the turtle step         */
/* is equal to an integer number of pixels.                    */
/*                                                             */
/* Characters listed after the "ignore" keyword                */
/* are considered as nonexistant while context matching.       */
/* Usually these characters represent geometric information    */
/* which is irrelevent from the viewpoint of interaction       */
/* between the components of the modelled structure.           */
/*                                                             */
/* The separators < and  > must be present in each production. */
/* The empty strings are denoted by the symbol *. All components */
/* of a production (including the strict predecessor) may contain */
/* several letters.  For example, AB < CDE > FG --> A[B]A      */
/* is a valid production which replaces substring CDE by A[B]A. */
/* The context lengths may vary from one production to another */
/* within a particular production set.                         */
/* Brackets can appear only in the production successor.      */
/***************************************************************/

#include <stdio.h>
#include "generate.h"

/***************************************************************/
/* "main" organizes computation.                               */
/***************************************************************/

main(argc, argv)
int argc;
char *argv[];
{
    char *title, *ctop();
    char *string, *Derive();
    char axiom[MAXAXIOM];
    static char ignore[MAXCHARS];
    Production prodSet[MAXPROD];
    Parameter p;

    if (argc != 2) {
        printf("Usage: %s filename\n", argv[0]);
        exit(1);
    }
    p.filename = argv[1];
    input(axiom, ignore, prodSet, &p);        /* read the input file  */
    string = Derive(axiom, ignore, prodSet, p.n); /* generate         */
    interpret(string, &p);                    /* interpret            */
}

/***************************************************************/
```

```
/* Function "input" reads the input file.                              */
/**********************************************************************/

input(axiom, ignore, prodSet, pPtr)
char axiom[];
char ignore[];
Production prodSet[];
Parameter *pPtr;
{
    FILE *fp, *fopen();
    char lcontext[MAXAXIOM], rcontext[MAXAXIOM], predecessor[MAXAXIOM];
    char successor[MAXAXIOM];
    char ignore_buf[MAXCHARS];
    char *inpStr; /* inpStr will contain the contexts, the strict
                     predecessors and the successors of all
                     productions, separated by null characters      */
    char *malloc();
    int i;

    if ((fp = fopen(pPtr->filename, "r")) == NULL) {
        printf("Can't open %s\n", pPtr->filename);
        exit(1);
    }
    if ((inpStr = malloc(MAXSTR)) == NULL) {
        printf("Can't allocate inpStr\n");
        exit(1);
    }

    *inpStr++ = '\0'; /* Start inpStr with the null character
            needed to terminate the search for the left context   */

    fscanf(fp, "derivation length: %d\n", &pPtr->n);
    fscanf(fp, "angle factor: %d\n", &pPtr->angle);
    fscanf(fp, "scale factor: %d\n", &pPtr->scale);
    fscanf(fp, "axiom: %s\n", axiom);
    fscanf(fp, "ignore: %s", ignore_buf);

    /* Set flags corresponding to the ignored characters.             */

    for (i=0; i<MAXCHARS; i++)
        ignore[i] = 0;
    for (i=0; ignore_buf[i] !=0; i++) {
        ignore[ignore_buf[i]] = 1;
    }

    /* Read productions and enter them into the "prodSet" structure */

    for (i=0; 1; i++, prodSet++) {
        prodSet->pred = inpStr;
```

```
        if (fscanf(fp, "%s < %s > %s --> %s",
                lcontext, predecessor, rcontext, successor) == EOF)
            break;
        enter(lcontext, &prodSet->lConLen, &prodSet->lCon, &inpStr);
        enter(predecessor, &prodSet->predLen, &prodSet->pred, &inpStr);
        enter(rcontext, &prodSet->rConLen, &prodSet->rCon,  &inpStr);
        enter(successor, &prodSet->succLen, &prodSet->succ, &inpStr);
    };

    prodSet->predLen = 0;
    printf("%d productions read\n", i);
}

/*****************************************************************/
/* Function "enter" is used to copy "string" to "inpStr" and fill  */
/* the fields in a "prodSet" structure.                             */
/*****************************************************************/

enter(string, lenPtr, prodStrHandle, inpStrHandle)
char *string;
int *lenPtr;
char **prodStrHandle, **inpStrHandle;
{
    if (*string == '*') {
        *lenPtr = 0;
        **inpStrHandle = '\0';
    }
    else {
        *lenPtr = strlen(string);
        strcpy(*inpStrHandle, string);
    }
    *prodStrHandle = *inpStrHandle;
    *inpStrHandle += *lenPtr + 1;
}

/*****************************************************************/
/* Given an axiom, a set of productions and a derivation length,   */
/* function "Derive" creates the generated string and returns a    */
/* pointer to it. The parallel operation of an L-system is         */
/* simulated as follows. First, string1 (containing the axiom) is  */
/* scanned from the left to the right. Its consecutive substrings  */
/* are matched with the predecessors of productions. The           */
/* appropriate successors are appended to the end of string2.      */
/* After all of string1 is scanned, string2 becomes string 1.      */
/* The process is repeated until the desired derivation length     */
/* is achieved.                                                    */
/*****************************************************************/

char *Derive(axiom, ignore, prodSet, n)
```

```
char axiom[];
char ignore[];
Production prodSet[];
int n;
{
    char *curPtr, *nextPtr, *tempPtr, *limPtr;
    char *string1, *string2;
    int i;
    Production *FindProd();

    if ((string1 = malloc(MAXSTR)) == NULL) {
        printf("pfg: can't allocate string1\n");
        exit(1);
    }
    if ((string2 = malloc(MAXSTR)) == NULL) {
        printf("pfg: can't allocate string2\n");
        exit(1);
    }
    for (i=0; i < MAXSTR; i++)
        *(string1+i) = *(string2+i) = '\0';
    ++string1;        /* start with '\0' for proper context handling */
    ++string2;

    strcpy(string1, axiom);
    limPtr = string2 + MAXSTR - MAXAXIOM;
    for (i=1; i<=n; i++) {
        printf("Computing derivation step %d\n", i);
        curPtr = string1;
        nextPtr = string2;
        while (*curPtr != '\0') {
            ApplyProd(&curPtr, &nextPtr,
                                    FindProd(curPtr, prodSet, ignore));
            if(nextPtr > limPtr) {
                printf("String too long");
                exit(1);
            }
            *nextPtr = '\0';
        }
    tempPtr = string1;
    string1 = string2;
    string2 = tempPtr;
    }
    return(string1);
}

/******************************************************************/
/* Copy the successor of the production *prodPtr starting at     */
/* location *nextHandle. Update curHandle and nextHandle.         */
/******************************************************************/
```

```
ApplyProd(curHandle, nextHandle, prodPtr)
char **curHandle, **nextHandle;
Production *prodPtr;
{
    if (prodPtr != NULL) {
        strcpy(*nextHandle, prodPtr->succ);
        *curHandle += prodPtr->predLen;
        *nextHandle += prodPtr->succLen;
    }
    else {
        **nextHandle = **curHandle;
        ++(*nextHandle);
        ++(*curHandle);
    }
}

/*******************************************************************/
/* Given a pointer to a string and a set of productions, return the */
/* pointer to the first applicable production or NULL if no         */
/* matching production can be found. The set of productions is      */
/* scanned in the same order in which productions are listed in     */
/* the input file.                                                  */
/*******************************************************************/

Production *FindProd(curPtr, prodSet, ignore)
char *curPtr;
Production prodSet[];
char ignore[];
{
    while (prodSet->predLen != 0) {
        if (prefix(prodSet->pred, curPtr) ||
            rcondiff(prodSet->rCon, curPtr + prodSet->predLen, ignore) ||
            lcondiff(prodSet->lCon + prodSet->lConLen - 1,
                                        curPtr - 1, ignore))
            ++prodSet;
        else
            return(prodSet);
    }
    /* Predecessor not found */
    return(NULL);
}

/*******************************************************************/
/* Check whether string s1 is a prefix of string s2.                */
/*******************************************************************/

prefix(s1, s2)
char *s1, *s2;
```

```
{
    while (*s1 != '\0')
        if (*s2++ != *s1++)
            return(1);
    return(0);
}

/********************************************************************/
/* Check whether string s1 matches s2 as the right context.         */
/* Ignore specified symbols and skip over branches.                 */
/********************************************************************/

rcondiff(s1, s2, ignore)
char *s1, *s2;
char ignore[];
{
    char *skipright();

    while(1) {
        if (*s1 == '\0')
            return(0);
        if(ignore[*s2])
            s2++;
        else if (*s2 == '[')
            s2 = skipright(s2+1) +1;
        else  if (*s1++ != *s2++)
                return(1);
    }
}

/********************************************************************/
/* Skip over a branch while searching for the right context. The    */
/* branch may contain subbranches.                                  */
/********************************************************************/

char *skipright(s)
char *s;
{
    int level = 0;

    while (*s != '\0') {
        switch (*s) {
        case '[':
            ++level;
            break;
        case ']':
            if(level == 0)
                return(s);
            else
```

```
                --level;
            break;
        default:
            break;
        }
        s++;
    }
    return(s);
}

/*******************************************************************/
/* Check whether string s1 matches s2 as the left context.         */
/* Ignore specified symbols and skip over branches.  The parent    */
/* branch belongs to the left context of a child branch.           */
/*******************************************************************/

lcondiff(s1, s2, ignore)
char *s1, *s2;
char ignore[];
{
    char *skipleft();

    while(1) {
        if(*s1 == '\0')
            return(0);
        if(ignore[*s2] || (*s2 == '['))
            s2--;
        else if (*s2 == ']')
            s2 = skipleft(s2);
        else {
            if (*s1-- != *s2--)
                return(1);
        }
    }
}

/*******************************************************************/
/* Skip over a branch while searching for the left context. The    */
/* branch may contain subbranches.                                 */
/*******************************************************************/

char *skipleft(s)
char *s;
{
    int level = 0;

    s--;
    while(*s != '\0') {
        switch(*s) {
```

```
        case ']':
            ++level;
            break;
        case '[':
            if(level == 0)
                return(--s);
            else
                --level;
            break;
        default:
            break;
        }
        s--;
    }
    return(s);
}
```

A.3 interpret.h

```
#define MAXANGLE    40      /* maximum value of the angle factor    */
#define MAXSCALE    100     /* maximum value of the scale factor    */
#define TWO_PI      6.2831853
#define STACK_SIZE  40      /* maximum depth of branches            */
#define LEFT        2       /* window parameters                    */
#define TOP         42
#define RIGHT       510
#define BOTTOM      340

struct TURTLE {             /* turtle position and orientation      */
    double x;
    double y;
    short int dir;
    };

typedef struct TURTLE TURTLE;

struct PIXEL {              /* turtle position in screen            */
    short int h, v;         /* coordinates                          */
    };

typedef struct PIXEL PIXEL;

struct BOX {                /* the bounding box of the curve        */
    double xmin, xmax;
    double ymin, ymax;
    };

typedef struct BOX BOX;
```

A.4 interpret.c

```
#include <quickdraw.h>
#include <font.h>
#include <window.h>
#include <math.h>
#include "generate.h"
#include "interpret.h"

/*****************************************************************/
/* Create the graphics environment in which the curve will be    */
/* drawn. Find the bounding box of the curve, center it and draw */
/* using the specified parameters.                               */
/*****************************************************************/

interpret(string, pPtr)
char *string;
Parameter *pPtr;
{
    Rect boundsRect;        /* needed to open a window on the Mac... */
    WindowRecord wRecord;
    WindowPtr picture;
    short int inc;          /* the step size                         */
    PIXEL start;            /* starting position of the turtle       */
    BOX boundBox;           /* bounding box of the curve             */

        /* Check the value of the angle factor.                      */

    if (pPtr->angle > MAXANGLE) {
        printf("Angle factor too big (%d > %d)\n",
            pPtr->angle, MAXANGLE);
        exit(1);
    }

        /* Compute the bounding rectangle of the curve, assuming     */
        /* that the mouse starts at the point (0,0) and the step     */
        /* size is equal to 1.                                       */

    start.h = start.v = 0;
    inc = 1;
    draw(string, &start, inc, pPtr->angle, 0, &boundBox);

        /* Given the bounding rectangle and the size factor,         */
        /* center the figure in the window.                          */

    SetDrawParam(&start, &inc, &boundBox, pPtr->scale);

        /* Initialize the drawing environment.                       */
```

```
    InitGraf(&thePort);
    InitFonts();
    InitWindows();
    SetRect(&boundsRect, LEFT, TOP, RIGHT, BOTTOM);
    picture = NewWindow(&wRecord, &boundsRect, ctop(pPtr->filename),
        257, documentProc, (WindowPtr) 0, 257, 0L);
    OpenPort (thePort);
    SelectWindow (picture);
    SetPort(picture);
    HideCursor();

        /* Draw curve.                                                 */

    draw(string, &start, inc, pPtr->angle, 1, &boundBox);

    ClosePort(thePort);
}

/*********************************************************************/
/* Draw figure defined by the "string". The parameters have           */
/* the following meanings:                                             */
/*                                                                     */
/*  string      the string being interpreted                           */
/*  startPtr    initial position of the turtle                         */
/*  inc         step size                                              */
/*  angFac      the angle factor                                       */
/*  flag        0 - make all lines invisible, 1 - draw. The value      */
/*              of 0 is used when calculating the bounding             */
/*              rectangle.                                             */
/*  boxPtr      Pointer to the structure returning the coordinates     */
/*              of the bounding rectangle.                             */
/*                                                                     */
/* The following string symbols are interpreted by the turtle:        */
/*                      +  -  |  [  ]  F  f                            */
/*********************************************************************/

draw(string, startPtr, inc, angFac, flag, boxPtr)
char *string;
PIXEL *startPtr;
short int inc, angFac, flag;
BOX *boxPtr;
{
    double SI[MAXANGLE], CO[MAXANGLE];
    TURTLE tu, stack[STACK_SIZE], *stackPtr, *bottom, *top;
    char c;
    int i, halfangFac;
    double ang = -TWO_PI/4;
    char *str;
```

```
    str = string;

    /* Set stack limits.                                                  */

    stackPtr = bottom = stack;
    top = stack + STACK_SIZE - 1;

    /* Precalculate coordinates for turtle steps in all possible          */
    /* directions.                                                        */

    for(i=0; i<angFac; i++) {
        SI[i] = inc * sin(ang);
        CO[i] = inc * cos(ang);
        ang += TWO_PI/angFac;
    };

    halfangFac = angFac/2; /* needed to interpret the symbol |            */
    angFac--;              /* more convenient for comparisons             */


    /* Initialize the bounding rectangle and the starting position        */
    /* of the turtle for the purpose of bounding rectangle                */
    /* calculation.                                                       */

    boxPtr->xmin = boxPtr->xmax = tu.x = (double) (startPtr->h) + 0.5;
    boxPtr->ymin = boxPtr->ymax = tu.y = (double) (startPtr->v) + 0.5;
    tu.dir=0;

    /* Move the turtle to an appropriate starting position in             */
    /* order to center the final drawing.                                 */

    if (flag)
        MoveTo(startPtr->h, startPtr->v);

    /* Scan the string and interpret its consecutive symbols.             */

    while ((c = *str++) != '\0') {
        switch (c) {
        case '+':               /* Turn right.                            */
            if(tu.dir<angFac)
                ++tu.dir;
            else
                tu.dir=0;
            break;
        case '-':               /* Turn left.                             */
            if(tu.dir>0)
                --tu.dir;
            else
                tu.dir=angFac;
```

```
            break;
        case '|':                 /* Turn around.                      */
            if(tu.dir>=halfangFac)
                tu.dir-=halfangFac;
            else
                tu.dir+=halfangFac;
            break;
        case '[':                 /* Push current state on stack.      */
            if(stackPtr >= top) {
                printf("Too many [\n");
                exit(1);
            }
            TurtleCopy(stackPtr, &tu);
            stackPtr++;
            break;
        case ']':                 /* Pop state from the stack.         */
            if(stackPtr <= bottom) {
                printf("Too many ]\n");
                exit(1);
            }
            --stackPtr;
            TurtleCopy(&tu, stackPtr);
            if (flag)
                MoveTo((short)(tu.x), (short)(tu.y));
            break;
        case 'F':                 /* Move forward and draw a line.     */
            tu.x += CO[tu.dir];
            tu.y += SI[tu.dir];
            if (flag)
                LineTo((short)(tu.x), (short)(tu.y));
            else
                BoxUpdate(&tu, boxPtr);
            break;
        case 'f':                 /* Move forward without drawing.     */
            tu.x += CO[tu.dir];
            tu.y += SI[tu.dir];
            if (flag)
                MoveTo((short)(tu.x), (short)(tu.y));
            else
                BoxUpdate(&tu, boxPtr);
            break;
        default:                  /* room for extensions               */
            break;
        }
    }
}

/*****************************************************************/
/* Function "TurtleCopy" copies structure "fromPtr" to "toPtr".     */
```

```
/******************************************************************/

TurtleCopy(toPtr, fromPtr)
TURTLE *toPtr, *fromPtr;
{
    toPtr->x = fromPtr->x;
    toPtr->y = fromPtr->y;
    toPtr->dir = fromPtr->dir;
}

/******************************************************************/
/* Check whether the turtle moves outside the current box. If it    */
/* does, adjust the box boundary.                                   */
/******************************************************************/

BoxUpdate(tuPtr, boxPtr)
TURTLE *tuPtr;
BOX *boxPtr;
{
    if (tuPtr->x < boxPtr->xmin)
        boxPtr->xmin = tuPtr->x;
    if (tuPtr->x > boxPtr->xmax)
        boxPtr->xmax = tuPtr->x;
    if (tuPtr->y < boxPtr->ymin)
        boxPtr->ymin = tuPtr->y;
    if (tuPtr->y > boxPtr->ymax)
        boxPtr->ymax = tuPtr->y;
}

/******************************************************************/
/* Set the starting point and the step increment, given the        */
/* bounding box and the scale factor.                               */
/******************************************************************/

SetDrawParam(startPtr, incPtr, boxPtr, scale)
PIXEL *startPtr;
short int *incPtr;
BOX *boxPtr;
int scale;
{
    double xscale, yscale, sc;

    xscale = (RIGHT - LEFT)/(boxPtr->xmax - boxPtr->xmin);
    yscale = (BOTTOM - TOP)/(boxPtr->ymax - boxPtr->ymin);

    if(xscale>yscale)
        sc = yscale;
    else
        sc = xscale;
```

```
    *incPtr = (int) floor ((double) ((sc * scale)/MAXSCALE));

    startPtr->h = (short)
        (RIGHT - LEFT - *incPtr * (boxPtr->xmin + boxPtr->xmax - 1.0))/2;
    startPtr->v = (short)
        (BOTTOM - TOP - *incPtr * (boxPtr->ymin + boxPtr->ymax - 1.0))/2;
}
```

Plate 1: A three-dimensional representation of *Lychnis coronaria* (*cf.* Figure 3.9).

Plate 2: A three-dimensional developmental model of *Mycelis muralis* (*cf.* Figure 3.15).

Plate 3: A lilac twig. Petals are represented by cubic patches as in the black-and-white twig of Figure 4.3. Leaves were synthesized using a developmental model.

Plate 4: An apple twig. Petals and leaves were modeled using cubic patches.